Plan for a Turbulent Future: Your Roadmap to Personal Resilience for a Changing Climate

RIVER PUBLISHERS SERIES IN ENERGY SUSTAINABILITY AND EFFICIENCY

Series Editor:

PEDRAM ASEF
Lecturer (Asst. Prof.) in Automotive Engineering,
University of Hertfordshire, UK

The "River Publishers Series in Sustainability and Efficiency" is a series of comprehensive academic and professional books which focus on theory and applications in sustainable and efficient energy solutions. The books serve as a multi-disciplinary resource linking sustainable energy and society, fulfilling the rapidly growing worldwide interest in energy solutions. All fields of possible sustainable energy solutions and applications are addressed, not only from a technical point of view, but also from economic, social, political, and financial aspects. Books published in the series include research monographs, edited volumes, handbooks and textbooks. They provide professionals, researchers, educators, and advanced students in the field with an invaluable insight into the latest research and developments.

Topics covered in the series include, but are not limited to:

- Sustainable energy development and management
- Alternate and renewable energies
- Energy conservation
- Energy efficiency
- Carbon reduction
- Environment

For a list of other books in this series, visit www.riverpublishers.com

Plan for a Turbulent Future: Your Roadmap to Personal Resilience for a Changing Climate

Remi Charron, Ph.D.

Consultant, Canada

LONDON AND NEW YORK

Published 2022 by River Publishers
River Publishers
Alsbjergvej 10, 9260 Gistrup, Denmark
www.riverpublishers.com

Distributed exclusively by Routledge
4 Park Square, Milton Park, Abingdon, Oxon OX14 4RN
605 Third Avenue, New York, NY 10017, USA

Plan for a Turbulent Future: Your Roadmap to Personal Resilience for a Changing Climate / by Remi Charron.

Routledge is an imprint of the Taylor & Francis Group, an informa business

ISBN 978-87-7022-768-1 (pbk)
ISBN 978-87-7022-755-1 (print)
ISBN 978-10-0080-280-1 (online)
ISBN 978-1-003-34921-1 (ebook master)

While every effort is made to provide dependable information, the publisher, authors, and editors cannot be held responsible for any errors or omissions.

Millions of lives and businesses have either been lost or significantly impacted by COVID-19. Yet, many are warning that climate change will be much more devastating over the coming decades. Reality is starting to set in. We are not going to achieve our global mitigation targets; we probably won't even come close. Individuals faced with this reality react differently, from willful ignorance to anxiety and depression, all normal reactions. If you believe in the science and understand the likelihood of our failure to meet our targets, you need to accept and plan for an unknown, challenging future. We may be individually powerless to stop climate change, but we certainly can act in ways that will help us better face the consequences. This book provides a structured approach to plan and prepare today for a world rocked by a turbulent climate.

Contents

Preface

Climate change has been a concern of mine since I was in elementary school and I read an article about how polar bears were starting to be impacted by the decline in sea ice. In university, I struggled like half of the current postsecondary students whose mental health is impacted by feelings of existential angst over climate change, job prospects and future economic stability. These struggles helped give me the motivation to do a Master's and Ph.D. in Building Engineering so that I could help fight climate change. However, the eco-anxiety and fears for the future haven't gone away. Like many experts, the more I learn, the more I worry about the ever-widening gap between what we need to do to tackle climate change versus what we are actually doing.

It isn't just climate change that keeps me up at night. Like the ozone hole, acid rain, the destruction of the Amazon rainforest, etc., climate change is just a symptom of a larger problem. We live on a finite planet, yet our whole socio-economic system is set up on a premise that we can have exponential growth go on forever. That we can keep extracting more and more resources, and using the earth as a giant dump for our waste forever. Where it's okay that a handful of super-rich individuals can be wealthier than the majority of humans on the planet. Where the disparity between the rich and poor keeps growing, year after year.

When I start thinking about all of the problems that we currently face and the fact that climate change is bound to make things much worse in the decades to come, I often get lost in despair. Although I struggle to find hope that society will get its act together in time to limit climate change, what does give me hope is to remind myself that societies have fallen in the past, but that humanity has carried on. Humans are resilient. We have faced much harsher conditions in the past and have thrived despite having much less technologies and knowledge than we have available today.

Resilience is the new buzz word. Everyone is talking about it. The Merriam-Webster dictionary defines it as "an ability to recover from or adjust easily to misfortune or change". There is a lot of research going on about how to make just about anything more resilient, from houses to cities, from individuals to businesses. Nearing the end of my Ph.D. in 2006, I started my first attempt to write a book about resilience. However, having kids and starting a new job put the idea of a book on the back-burner.

For the next couple of years, my focus wasn't on writing a book, but on doing research on how I could make myself and my family more resilient. Based on this research I developed a resilience plan for my family, which I started to implement in 2010. As described in the introduction, putting this plan in motion completely altered the path I had been on for my life up to that point. The thought of writing a book was pushed further aside.

Fast forward to 2020. Like many people, COVID 19 had me working from home where my kids were being homeschooled. My 12-year old daughter likes to write. In the past, she had started a number of novels but failed to finish any of them. As a homeschooling project, I challenged her to write a book from start to finish and agreed to write a book at the same time.

Given that it had been 10 years since I had developed our family resilience plan, and 14 years since I had first tried to write a book about resilience, I decided to write a book to help others in creating their own resilience plan. This resulting book combines the years of research I did on personal resilience with the 10 years of experience I have had to implement my own plan. I hope you will find it useful in starting your personal resilience journey.

List of Figures

1

Planning for a Turbulent Future

"For tomorrow belongs to the people who prepare for it today."

– African Proverb

Abstract

Chapter 1 starts with a story on how the author embarked on a journey to improve his family's resilience a decade ago. As with any major life change, things weren't always easy. Despite the hardship, the reader is encouraged to embark on their own personal journey to improve their resilience.

Keywords: resilience plan, climate, sustainability, doomer.

1.1 My Resilience Journey

Ten years ago, I had an excellent job as a Senior Researcher in Sustainable Housing for the Canadian government, with two young kids and a loving wife, living in a nice house in Ottawa. One day, I quit my job, sold the house and moved in with my in-laws in their farmhouse in the outskirts of Pemberton, British Columbia. Had I lost my mind? The jury is still out on that one. Essentially, my wife and I realized that our society was unsustainable. This was not a shocking, life-altering realization on its own. However, if you continue with that line of reasoning, the obvious conclusion is that something unsustainable cannot be sustained indefinitely. Working in sustainable housing, I realized that most of what we consider "green" or sustainable is not actually sustainable; it is just less harmful than current practice.

There is a whole sub-culture of people who think that society will collapse for a variety of different reasons, and these people are generally disregarded and called "doomers". A bunch of chicken little's running around saying that the sky is falling. From the doomers group, there is a sub-group called "preppers" and they are diligently preparing for the collapse of society. The problem is that society is not collapsing right now and there is only so long that you can live thinking the sky is falling when everyone else is out enjoying life under the beautiful blue sky. Similarly, it is hard to prepare for something when you don't know exactly what will happen, nor when it will happen.

Although my wife and I didn't think that society's collapse was imminent, we certainly believed then, and still to this day believe that our kids will live through a turbulent future. It will not be the nice stable society that we have experienced since World War 2. Many people believe this on some level to the extent that there is a growing movement of people who don't have kids because they don't want them to grow up in the world we have created for them. However, for those who do have kids, we somehow need to prepare them for what's in store for them.

How do you prepare them, without knowing exactly what the future holds, other than a planet that has had its climate significantly altered and whose resources are being depleted after decades of overconsumption? We tried to imagine what professions we were confident would continue to exist regardless of how society is doing: health care, farming, mechanics, etc.? In the end, we decided that raising our kids on a farm would provide them with valuable life skills that will always be useful. We asked our in-laws if we could move in with them on their beef and potato farm. They agreed as long as we would do our own thing, as they couldn't afford to hire us to work on their farm. In a few short weeks, we sold our house, I quit my job and we moved across the country.

Before leaving, I wrote a seven-page manifesto. I sent it to all my friends and extended family, explaining why we thought society was on a collision course and why we felt moving to Pemberton was the right thing to do. I still cringe at the thought of sending such a personal email. It is probably the last communication that most of those people have had from me, other than a Facebook post once in a while.

1.2 How Did it Go?

Telling you that the last 10 years have been great would be a lie. Within a few weeks of moving, I had a terrible case of seller's remorse. Within a few months, I was ready to move back to Ottawa in defeat. However, once I convinced Carrie, my wife, to move back to her family home in Pemberton, there was no way she was going to leave. It almost cost us our marriage and my sanity, but we managed to persevere through it and are now both happy with our lives. I am someone that likes to live life with no regrets. It has only been a couple of years that, given a chance, I would not go back to the past and change that decision to move my family to Pemberton.

Why was it so bad? It was hard on many levels. Although I anticipated it to be difficult to live in my in-law's house, it was harder than I had imagined. I am not sure why I underestimated it. I didn't think I would be able to move back in with my parents for more than a few weeks. Why would moving in indefinitely with my wife's parents be any better? I had accepted that living in close proximity all the time could be challenging, but when you are actually living together, sharing every meal, getting in each others' way, it can get overwhelming. Even though it was hard on me, I can only imagine what it was like for my in-laws, going from a quiet house with lots of room, to a crowded house with loud kids running around. The fact that we had differences of opinions on how kids should be raised and disciplined was a definite friction point. After a couple of years, my in-laws renovated the house to make a suite for themselves on the ground floor. Having our own space, only having one breakfast and one dinner together per week, has been essential in restoring some sanity in the house.

I certainly can't blame all the hardship on sharing a house with my in-laws. There were a lot of other challenging experiences. In our first year, Carrie and I started a market-garden selling fruits and vegetables at the farmer's market. Although I love my wife very much, I found working together very difficult. With having her grow up on a farm, I always felt like an employee, being told what to do and what I did wrong. It certainly didn't help that farming, it turns out, is much harder than it looks. Two engineers, one with a

Ph.D., working from 6 am to 11 pm, day in and day out to make less money than you would be working for minimum wage at a fast-food restaurant is demoralizing. You can't expect to make much money in the first year of any new business. However, I soon realized that most market gardeners are just scraping by, putting in insane hours.

After a year, I started doing some consulting to make more money for our family and carve out something of my own outside of the farm. Five years ago, without discussing it first with my wife, I applied to be the program coordinator and professor for a new Master of Science in Energy Management at the New York Institute of Technology's Vancouver campus. Lo-and-behold, I got the job. My wife was mad, both that I would have to leave her 4 days a week while I stayed and worked in Vancouver, and second, that I hadn't discussed it with her first. We once again had to adjust to a new reality. In the end, it turned out to be beneficial for our marriage, which was good for our family. I was happier and the space helped us reconnect after some difficult years.

Despite my challenging experience, I decided to write this book because I want to encourage other people to start on their path to preparing for the future that awaits us. You probably don't need to embark on such a drastic life-changing journey that my wife and I started on. This book lays out a plan that you can personalize that best reflects your situation. Most of the things outlined in this plan will benefit you in life, regardless of what happens, so you have nothing to lose.

1.3 Why Plan Now?

Many people found themselves unprepared for all the disruptions that COVID 19 caused. No one could have anticipated that 2020 would bring such disruption to their income, their supermarket, their social life, their family life, etc. Many people lost their jobs, businesses closed for good, marriages broke up, mental health deteriorated for millions. No one could have planned for this particular global pandemic. However, as with other disasters or hardships, faced with the same problems, some people seem to take everything in stride and others seem to fall apart. This has to do with how resilient people are. It is not surprising that there has been an increase in people wanting to learn how to become more resilient. Although it is never too late to become more resilient, it is more difficult to do in the middle of a crisis.

If someone had a crystal ball and could predict the future, it would be easy to prepare and plan for what life had in store for you. Unfortunately, no one can predict the future. Becoming more resilient requires you to anticipate and prepare for future events that can occur (earthquakes, floods, recessions, pandemics). More broadly, it is to prepare you to be able to face and persevere through unexpected events.

These days, we don't need a crystal ball to predict that climate change will bring hardship to our lives. As bad as the COVID 19 impacts have been, you don't have to look too far to find experts warning how climate change will cause more drastic implications in our lives in the coming decades. Not only will it increase the severity and the frequency of severe weather events bringing localized disasters, but it will also cause significant disruptions to the food supply, mass migrations and take a heavy economic toll. The cost of implementing sufficient global greenhouse gas (GHG) reductions will be enormous, whereas the cost of not reducing emissions can be unthinkable.

It is not only people that are worried. The insurance industry is becoming concerned about the increasing amount and cost of disasters they need to cover. As explained later in the book, with some of the climate disruptions becoming too frequent, the insurance industry has been pulling out and not providing coverage. It is not just the insurance industry that is worried. A recent survey of 500 C-level executives from global companies is showing that most industry leaders are becoming concerned with the effects of climate change, with 78% saying that climate change is a serious threat to job security in the next 5 years[1].

The industry executives aren't just worried about a changing climate. They are concerned about the increasing pressure they get from shareholders and regulators to disclose the financial impact of extreme weather and more stringent regulations. Total economic costs could exceed $1 trillion in the next 5 years from the impact to supply chains, rising insurance costs, and increasing regulation. When you factor in that many of the companies will

---

[1]Fujita, A., (2020) 74% of executives say climate change poses risks to job security: survey, yahoo!finance, November 14, 2020. Available at: https://apple.news/AQnQ3MOOCTOqkFR 2oMleZSg (Accessed on: Dec. 29, 21).

be recovering from COVID 19 related losses for years to come, it can lead to job losses. This means that you will not only be facing more direct impacts of climate change through more natural disasters, but the effect of these disasters and the effort to reduce climate change will also start to have more indirect impacts on your life as well.

There is still a lot of uncertainty regarding if and how climate change will impact you personally in the coming years. Why should you start planning now to become more resilient to prepare for climate change impacts and other potential disruptions to your life with all of this uncertainty? Becoming more resilient is like investing for retirement; the more time you have to prepare, the more you can spread your level of effort.

1.4 Why This Plan?

Ten years ago, before I embarked on my crazy journey towards resilience, I developed a plan in an Excel worksheet to help prepare my family and me for our uncertain future. The plan's structure was based on many hours of research that I did in this area at the time. Since then, I have set my plan in motion and have lived through its implementation. A decade later, I have taken my original Excel worksheet and have made some modifications based on what I have learned since then, both at home and in my consulting work. I also reviewed more literature on personal resilience, which has expanded considerably over the years. The result is a more holistic personal resilience plan, which is more comprehensive than the others I have found.

You will find that most of the plan elements are things that you already know you should be doing: exercise, eat healthily, meditate, keep learning, embrace friends and family, stay out of debt, save for the future and be prepared for emergencies. There are many things we know we should do, but we rarely take the time to plan how and when we will do those things. Increasing your resilience will better prepare you to face life's unexpected and expected future challenges; with the added side-benefits of increased happiness and personal well-being.

2

Climate Targets Missed – So What?

"Extinction is the rule. Survival is the exception."

– Carl Sagan

Abstract

Chapter 2 explains that it is time to accept that we will not meet our aspirational goal of keeping the global increase in temperature below our 2 °C target and that we need to plan accordingly. That doesn't mean we should give up on trying to limit our impact on the climate. However, passing this safe operating limit increases the likelihood that we reach tipping points that will make climate change go from bad to worse.

Key Words: population growth, exponential growth, Jevon's paradox, tipping points, positive feedback, runaway global warming.

It is time for us to accept that we will not meet our aspirational goal of keeping the global increase in temperature to 1.5 °C, nor our 2 °C target. Just because it is still mathematically possible that globally we can still achieve these targets, it is not in the realm of possibilities given the level of coordinated effort and sacrifices that it would take. This does not mean that it is time to give up on trying to reduce GHG emissions. If anything, we need to redouble our efforts and do as much as possible to reduce emissions. It means that we need to take a serious look at what missing the 2 °C target means. Wasn't that the safe limit?

2.1 Why We Will Not Achieve Our Targets

We have already successfully addressed major environmental challenges. National action combined with international collaboration helped reduce acid rain and stopped the use of ozone-depleting chemicals that damage the ozone layer. Why can't we get our act together and address climate change? There are a number of factors that make it a more challenging problem to tackle. The main issue is that the global economy is very highly correlated to global GHG emissions. Energy is the fuel in the tank that drives our economy. Energy efficiency and renewable energy can help reduce the correlation between GHG emissions and the economy, however not at the scale of reduction that is needed. Several factors make it all the more difficult:

Global Population: Although population growth peaked in 1968 at 2.1% per year and has decreased to 1.1% per year in 2019, with a projected decrease to 0.1% per year by 2100, the global population is expected to go from 7.7 billion in 2019, up to 9.7 billion in 2050 and 10.9 billion in 2100.[1] In the next 30 years, we will add 2 billion people, the equivalent to our total global population of 1928. In 30 years, our population will increase by 25%, yet we are expected to reduce our net GHG emissions by 100% in that same time.

Embodied Carbon: The scale and speed at which we need to convert the entire global economy from being primarily based on fossil fuels to one based on renewable energy are mind-boggling. Building all of the electric cars, wind turbines, solar panels, etc., that we need requires an enormous amount of resources. Although these low carbon technologies might not emit GHGs when they are operating, there can be significant GHGs emitted to get the products built, from mining resources to manufacturing.

Keeping Up With Growth: Not only do we expect to increase our global population by 25% in 30 years, but we also expect the economy to grow by 140% over that period.[2] Growing our economy that much will increase our overall energy demand.

---

[1]Roser, M., (2019) Future Population Growth, Our World in Data, November 2019. Available at: https://ourworldindata.org/future-population-growth (Accessed on: Dec. 31, 2021).

[2]Historic and projected GDP in the EU, the US, the BRIICS countries and other countries, European Environment Agency, February 23, 2017

Jevon's Paradox: Jevon's paradox, otherwise known as the rebound effect, states that as we become more efficient at consuming energy, we tend to increase consumption. Take, for example, a family that trades in their gas-guzzling SUV for a fuel-efficient sedan. Since it costs less to operate, they might decide to drive further (direct rebound) or they could take those savings and take a family trip overseas, emitting a lot of emissions with their flights (indirect rebound). Essentially it boils down to the fact that when we save money on energy, those savings are generally reinvested into the economy and that economic activity comes with GHG emissions. Although our measures may reduce emissions, the net reduction is generally not as much as is assumed.

All of these factors combined can begin to explain how we keep emitting record levels of GHG globally, despite the astounding progress we have made in the last 20 years in terms of energy efficiency and renewable energy development. In terms of energy efficiency, over the previous 20 years, we have reduced the amount of global energy consumption per unit of GDP at a rate of 1.6% per year, for a total reduction of 26% in 20 years.[3] However, during those same years, energy consumption has grown by 2%per year or 40% overall. Simply put, our global economy has grown faster than the gains we have made with energy efficiency. The good news is that renewable energy is making up more and more energy consumption growth. In 2000, renewable energy accounted for around 20% of the added new annual energy capacity, whereas by 2018, renewable energy made up over 60% of added energy capacity.[4]

There are so many factors to keep track of that it can become easy to get lost in all of this data. It is easy to cherry-pick the good news to show how much progress we have made, or to look at all of the bad news and argue that we haven't been doing anything. There is only one indicator that matters: the concentration of greenhouse gases in the atmosphere. Most of the other parameters rely on reporting, whereas measuring concentrations of CO_2 and other greenhouse gases in the atmosphere is much less prone to errors or manipulation. As discussed in the next section, we are at and approaching specific tipping points where the planet's natural processes could start emitting more GHGs than humans (e.g. permafrost thawing), making it

---

[3]Global Energy Statistical Yearbook, Enerdata, https://yearbook.enerdata.net/total-energy /world-energy-intensity-gdp-data.html (Accessed on: Dec. 31, 2021)

[4]Renewable Capacity Statistics 2019, International Renewable Energy Agency, March 2019

difficult to track human related emissions combined with climate change related emissions.

There are a few charts that we can look at to give a perspective on our progress. The first (Figure 2.1a) displays the global CO_2 concentrations with the dates and locations of the annual international meetings that have taken place to tackle climate change.[5] Figure 2.1b shows an even steeper increase for methane (CH_4), a more potent greenhouse gas. No need to look at all of the (broken) promises and commitments, policies or emissions data. This chart says it all. The best indicator of future behavior is past behavior. It seems that almost every year we break a record in the increase in CO_2 and CH_4 emissions, despite the increasing amount of efforts implemented globally to tackle climate change.

Where does the CO_2 concentration need to go to reach our target of a 2 °C temperature increase? Concentrations would need to peak at no more than 450 parts per million for the 2 °C limit or 430 ppm for 1.5 °C.[6] To achieve that, global emissions need to peak now and start to decline rapidly, down to a low enough level that we can actually suck more carbon out of the air than we produce within 30 to 50 years. Global GHG reduction commitments are nowhere near where they need to be to achieve these targets and based on the past, countries generally do not achieve the level of reductions that they commit to.

The Climate Action Tracker is an independent scientific analysis that tracks government climate action and estimates what they would achieve in terms of global warming.[7] Figure 2.2 presents their 2021 update. Current pledges and targets would not see emissions peak before after 2030, and actual policies would see emissions peak between 2030 and 2050. This is much less than required. Roughly speaking, to achieve our 2 °C target, global emissions will need to reduce by around 20% by 2030, whereas to achieve our 1.5 °C target, we need a reduction of about 40%.

---

[5]Saxifrage, B. (2018) CO2 vs the COPs, Canada's National Observer, Dec. 12, 2018. Available at: www.nationalobserver.com/2018/12/12/analysis/co2-vs-cops (Accessed on: Dec. 31, 2021)

[6]Kemp, J. (2019) Climate change targets are slipping out of reach, Reuters, Apr. 16, 2019, Available at: www.reuters.com/article/energy-climatechange-kemp-idUSL5N21Y4A0 (Accessed on: Dec. 31, 2021)

[7]Global update: Paris Agreement Turning Point, Climate Action Tracker, Dec. 1, 2020, Available at: https://climateactiontracker.org/press/global-update-paris-agreement-turning-point/

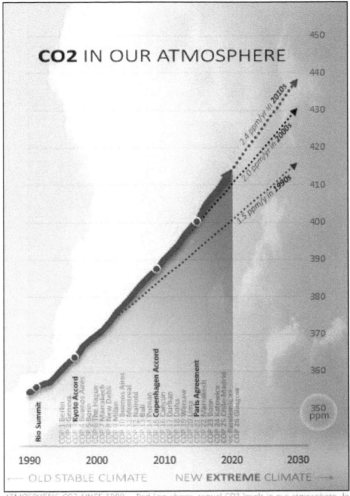

Figure 2.1a CO_2 in the atmosphere keeps increasing despite many international meetings to fight climate change.

To put it in perspective, to achieve the 2 °C target, we would need to eliminate the equivalent of all oil-related emissions before 2030. Achieving the 1.5 °C target would be equivalent to roughly all oil and coal emissions in that same period. These reductions would need to be done while keeping other emissions steady. We achieved around a 17% to 25% reduction in

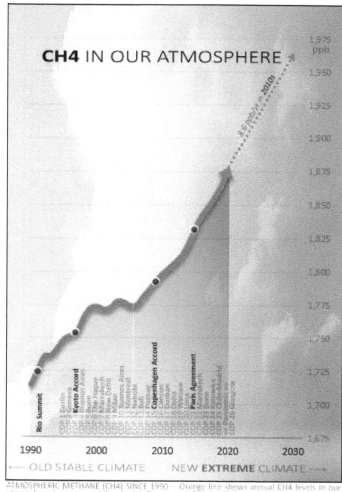

CH4 IN OUR ATMOSPHERE

Figure 2.1b Methane in the atmosphere keeps increasing despite many international meetings to fight climate change.

emissions at the peak of COVID restrictions in April 2020.[8] This was when most people were self-isolating at home reducing transportation emissions significantly. Emissions bounced back rapidly, where at week 40 of the

---

[8]Quéré, C.L., et al., Temporary reduction in daily global CO_2 emissions during the COVID-19 forced confinement, *Nature Climate Change* **10**, 647-653, May 19, 2020

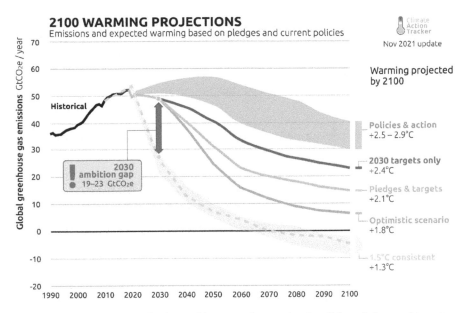

Figure 2.2 Warming projections with current international policies, pledges and targets.

pandemic, emissions were back to where they were the previous year.[9] It is not pessimistic to say that we will not achieve our 2 °C target; it is merely realistic.

Politicians tend to be dishonest in terms of the type of actions that would be required to achieve these targets. We have examples like Canadian Prime Minister Trudeau advocating for the 1.5 °C targets at the global climate meetings, yet turning around and buying the Trans Mountain pipeline for billions of dollars, committing Canada to decades of increased oil production levels. Trudeau was quoted as saying: "No country would find 173 billion barrels of oil in the ground and leave them there", which is precisely what will be required in order to achieve our global mitigation targets.

Globally we need to stop adding new fossil fuel capacity. Not only that, achieving our targets would mean that we actually need to start phasing-out fossil fuel production. When the international community starts to implement a planned phase-out of fossil fuels, this will indicate that they are

---

[9]Tollefson, J., COVID curbed carbon emissions in 2020 - but not by much, *Nature* **589**, 343, 2020

becoming more serious. They would have to first agree on the maximum amount of fossil fuels extracted by type (coal, oil and natural gas) and from which country and then decide how much each country is allowed to use. It is much easier to track and implement limits on fossil fuel extraction than to try to manage the billions of end-uses of these fuels around the world.

In 2017, a group of less developed countries asked for international climate negotiations to start phasing-out fossil fuels:[10]

> *"an increase in ambition by all countries to put us on track to limit the global temperature increase to 1.5 degrees Celsius by strengthening our national contributions, managing a phase-out of fossil fuels, promoting renewable energy and implementing the most ambitious climate action."*

This initial call has been growing, with over 150 legislators from 30 countries that have now endorsed a Fossil-fuel Nonproliferation Treaty.

2.2 Why Does it Matter

So we are not going to achieve our 2 °C target, how will that impact your life? If the climate behaved in a rational, linear way and it was distributed evenly across the world, warming of 2 °C might not be that bad. Unfortunately, that's not how it works. There are thresholds or tipping points, after which the impacts stop being proportional to the warming and we can get drastic, irreversible changes occurring. One reason for agreeing to limit warming to 1.5 °C or 2 °C was to prevent these drastic, irreversible changes.

When the Intergovernmental Panel on Climate Change (IPCC) introduced the idea of tipping points 2 decades ago, they assumed it would take warming of over 5 °C to trigger them.[11] However, their latest findings

---

[10]Fossil Fuel Non-Proliferation Treaty, Available at: https://fossilfueltreaty.org/ (Accessed on: Dec. 31, 2021)

[11]Lenton, T., et al., Climate tipping points - too risky to bet against, *Nature* **575**, 592-595, 2019

suggest that tipping points could be exceeded with just 1 to 2 °C of warming. It is not surprising, the IPCC process is very political and requires consensus, so to be accepted, it relies on conservative assumptions. Climate change is scary on its own, as we are already seeing more intense hurricanes, tornadoes, heat waves, etc., killing thousands and causing billions in damages every year. Triggering these tipping points will put the impacts of climate change on another level. Given the magnitude of their potential impact and the increasing likelihood that some of these will occur, I find it surprising how few people know about them. I will give a brief description of some of these potential impacts.

2.2.1 Shutdown of the Atlantic Meridional Overturning Circulation

The ocean has a current that brings warm salty water from the Gulf of Mexico to the North Atlantic. As the Greenland ice sheet melts, it introduces fresh water into the North Atlantic, which can slowdown or altogether stop the Gulf flow. It has already weakened by around 15% since the middle of the 20th century.[12] A shutdown would result in widespread cooling around the whole of the northern hemisphere, but particularly around Western Europe and the east coast of North America by as much as 5 °C. With the heat staying near the equator, rather than moving north, it would have other drastic impacts, altering weather patterns, damaging the Amazon rainforest and shrinking the West Antarctic ice sheet, to name a few.

2.2.2 Loss of Arctic and Antarctic Ice Sheets

The West Antarctic ice sheet holds enough frozen water to raise global sea levels by around 3.3 meters (11 ft).[13] It is particularly vulnerable because it sits on bedrock that is mostly below sea level and in contact with ocean heat, making it susceptible to rapid ice loss. The East Antarctic ice sheet might be similarly unstable and could add another 3 to 4 m (10 to 13 ft) to sea levels

---

[12]Potsdam Institute for Climate Impact Research (PIK). "Gulf Stream System at its weakest in over a millennium." ScienceDaily., 25 February 2021

[13]Lenton, T., et al., Climate tipping points – too risky to bet against, *Nature* **575**, 592-595, 2019

on timescales beyond a century. The Greenland ice sheet is melting at an accelerating rate and could add a further 7 m (22 ft) to sea level over thousands of years, which might be irreversible at 1.5 °C of warming. This is 13 to 14 m (42.5 to 45 ft) over hundreds or thousands of years. Geological records show that a sea level rise of 5 m (16 ft) in a century is extreme, but the rate of climate forcing is now more significant than any known natural forcing from the past.[14] The same study found that sea level was 6 to 9 m (20 to 29.5 ft) higher in the Eemian interglacial period, a time that was no more than a few tenths of a degree warmer than today.

2.2.3 Amazon Rainforest Dieback

Research has shown that reducing the amount of rainfall or the amount of forest can shift the Amazon climate into a drier state that could no longer support a rainforest. Increasing temperatures will lead to a decline in rainfall. At the same time, higher concentrations of CO_2 cause plants to lose less water and less transpiration means less water going back into the atmosphere. These two climate change impacts are compounded by deforestation. Fewer trees represent less evapotranspiration and less moisture entering the atmosphere. A tipping point will be reached when the rainforest can no longer sustain itself and transition to a savannah – a drier ecosystem dominated by open grasslands with few trees. The tipping point will depend both on the rising temperature and the amount of deforestation. Without global warming, the tipping point would be reached at around 40% deforestation. The dieback of the Amazon would lead to further global warming with the release of CO_2 from dying trees and a reduction in the amount of CO_2 absorbed annually by the forest.

2.2.4 West African and Indian Monsoon Shift

It isn't easy to simulate the monsoon system in climate models because various interrelated factors, including circulation, temperature and topography, come into play. Monsoons can switch from a wet state with lots

---

[14]Hansen, J., et al., Ice melt, sea level rise and superstorms: evidence from paleoclimate data, climate modeling, and modern observations that 2 °C global warming could be dangerous, *Atmos. Chem. Phys.* **16**, 3761-3812, 2016

of rain to dry states with little rain. There can also be shifts as to where the rain falls. The West African monsoon can shift southwards, causing droughts, amplified by climate change where drier conditions see less vegetation growth, a reduction in evapotranspiration and even less rainfall. In the past, the switch from a wet monsoon to a dry period was very abrupt, occurring within decades to centuries. Any change in the monsoon can be devastating to large numbers of people. With over a billion inhabitants, India receives around 70% of its annual rainfall during the monsoon season, with some areas as much as 90%.

2.2.5 Boreal Forest Shift

Boreal forests are the largest ecosystem on Earth's land surface and account for 30% of the world's forests. The boreal forests are seeing approximately twice as much warming as the global average. Increasingly warm summers are becoming too hot for the currently dominant tree species, increasing the trees' vulnerability to disease, decreasing reproduction rates and causing more frequent fires, all causing significantly higher tree mortality. Large or frequent fires coupled with climate change could make the forest incapable of regenerating and transforming into sparsely wooded or grassland ecosystems. This would, in turn, lead to more regional warming and increased fire frequencies, thus starting a positive feedback mechanism.

2.2.6 Impacts on our Ocean Wildlife

The oceans have been limiting the amount of climate change we have experienced by absorbing vast quantities of CO_2. However, this has been increasing the acidity of the ocean and making carbonate ions less abundant. Decreases in carbonate ions can make building and maintaining shells and other calcium carbonate structures difficult for calcifying organisms such as oysters, clams, sea urchins, shallow water corals, deep-sea corals and calcareous plankton.[15] Acidic oceans can even start to dissolve these carbonate structures. A recent study found that lower pH levels in the Pacific

[15]What is Ocean Acidification?, National Oceanic and Atmospheric Administration, Feb. 26, 2021, Available at: https://oceanservice.noaa.gov/facts/acidification.html, (Accessed on: Dec. 31, 2021)

Ocean are already causing parts of Dungeness crab shells to dissolve, damaging their sensory organs. The researchers found dissolution impacts to the crab larvae that were not expected to occur until much later in this century.[16] In addition to impacting calcifying organisms, more acidic oceans can also impact certain fish's ability to detect predators.

Impacts on fish are not limited to ocean acidification. Warming waters have less oxygen, making it harder for fish to breathe. Compounding the problem is that warming, low-oxygen waters also increase the amount of oxygen needed by fish.[17] Older, larger fish are more impacted because they have less gill area relative to their mass compared to younger fish. To compensate for the decreasing oxygen levels, fish adapt by either growing less, moving towards the poles or going deeper. The impacts of ocean acidification and the widespread movement of fish from the lower oxygen levels will have untold consequences on ecosystems, let alone impacting our global fisheries, which are already struggling with over-exploitation.

2.2.7 Permafrost and Methane Hydrates

To me, the scariest of the tipping points is the carbon locked up in the frozen permafrost on land and the large methane hydrates in the ocean. Globally, we emit the equivalent of around 50 billion tons of CO_2 each year into the atmosphere. The Earth's atmosphere already contains about 850 billion tons of carbon. Permafrost holds approximately 1400 billion tons of carbon, and methane hydrates have in the range of 1000 to 5000 billion tons. Both show signs that they are starting to thaw, releasing the previously trapped gases into the atmosphere.

The permafrost is thawing much faster than predicted. Models assumed that permafrost would thaw gradually from the surface downwards.[18] The problem is that the frozen soil physically holds the landscape together, and when it thaws, it can collapse suddenly as pockets of ice within it melt.

---

[16]Wray, M. (2020) The Pacific Ocean is now so acidic, Dungeness crab shells are dissolving, Global News, Jan. 28, 2020. Available at: https://globalnews.ca/news/6472038/ocean-acidity -dissolves-crab-shells/ (Accessed on: Dec. 31, 2021)

[17]University of British Columbia. "Theory explains biological reasons that force fish to move poleward." ScienceDaily. ScienceDaily, Oct. 28 2019.

[18]Turetsky, M., et al., Permafrost collapse is accelerating carbon release, *Nature* **569**, 32-34, 2019

Instead of the predicted few centimeters per year, several meters can collapse within days or weeks. The land can sink and be inundated by swelling lakes and wetlands.

It gets worse; scientists did not account for the behavior of the pesky beavers that are moving in.[19] Beavers seem to like damming streams in lake basin areas underlain by ice-rich permafrost. The pooled water warms the frozen soil, causing ice to melt and the ground to sink, deepening the ponds storing even more water and releasing more carbon.

Another mechanism accelerating permafrost melt has been noticed in Siberia after people discovered massive cylindrical sinkholes, in the range of 30 meters (100 ft) deep and 20 meters (66 ft) wide. The first sinkhole was discovered in 2014. Since then, 17 other holes have been found. They suspect that these are caused when pockets of methane gases trapped beneath the surface explode. According to Siberian Times, a review of satellite data revealed as many as 7,000 gas-filled 'bubbles' that could burst into more giant sinkholes.[20]

Science hasn't caught up with the reality of how permafrost behaves in a warming world. If the permafrost melt rate were to reach 3.5% per year, this would be equivalent to all of the GHGs that are emitted each year by humans. If we allow that to happen, we would have runaway climate change as no emission reductions could stop the climate from getting warmer. Maybe 3.5% is a high estimate. However, this doesn't factor in the potential melting methane hydrates in the ocean that contain even more carbon. Some research suggests that methane releases from hydrates could happen in as much as 50 billion tons released abruptly, equivalent to one year of human-generated global GHG emissions. As the science on hydrates evolves, it seems that we may have been underestimating this problem.

Countless other large irreversible impacts will occur as a result of our continued warming. With 2 °C of warming, 99% of tropical corals are projected to be lost. There are 800 million people who depend on melt water

---

[19]Chrobak, U., Beavers might be making the Arctic melt even faster, Popular Science, Jul. 1, 2020, Available at: www.popsci.com/story/environment/beavers-tundra-permafrost-melt/, (Accessed on: Dec. 31, 2021)

[20]7,000 underground gas bubbles poised to 'explode' in Arctic, The Siberian Times, Mar. 20, 2017, Available at: https://siberiantimes.com/science/casestudy/news/n0905-7000-underground-gas-bubbles-poised-to-explode-in-arctic/

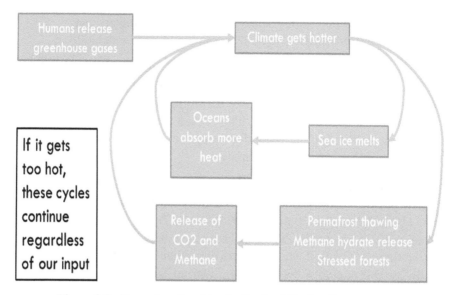

Figure 2.3   Examples of positive feedback within the climate system.

from the thousands of glaciers in the high mountains of Asia. Even if we were able to keep the temperature increases to 1.5 °C, over a third of the glaciers could be gone by the end of the century.[21]

A number of these tipping points could act as falling dominos, with one triggering another, which then triggers another. This is because many of these climate systems act with a positive feedback cycle (see Figure 2.3). The most obvious one is the melting of permafrost or hydrates. Once it starts to melt, it releases greenhouse gases, which causes more warming and more rapid melting of these vast stores of carbon. Melting sea ice is another prime example. Instead of ice that reflects sunlight, the oceans absorb sunlight. As more sea ice is lost, the seas absorb more solar radiation, becoming hotter and melting even more ice. Research looking at 30 different systems with potential tipping points found positive feedback links for 45% of possible interactions.[22] These would increase the level of global warming, likely

---

[21]Jordans, F., Asia's glaciers face massive melt from global warming, PHYS.ORG, Sep. 13, 2017, Available at: https://phys.org/news/2017-09-asia-glaciers-massive-global.html, (Accessed on: Dec. 31, 2021)

[22]Lenton, T., et al., Climate tipping points - too risky to bet against, *Nature* **575**, 592-595, 2019

triggering other tipping points. All of these could turn the planet into its less habitable, "hothouse" climate state.

2.3 Other Problems to Worry About

You might be unwilling to accept defeat and start planning for further climate disruptions or it might be that you are not convinced that you and your family will experience any significant consequences as a result of climate change. Unfortunately, climate change is only a symptom of a larger problem. Even if a breakthrough technology was invented tomorrow that could replace fossil fuels inexpensively and quickly, society wouldn't be out of the woods quite yet. Our whole socio-economic model is dependent on continuous economic growth. Perpetual growth in a finite world is a recipe for disaster.

Bill Rees, a professor from the University of British Columbia, came up with the ecological footprint concept to try to estimate how much resources individuals, communities or whole countries and the world consumed. Based on his assessment using 2014 population and consumption levels, Rees found that it takes four to 10+ hectares of global average productive ecosystems (gha) to produce the bio resources consumed by the average high-income citizen.[23] The average human eco-footprint was 2.8 gha, yet the planet can only sustain around 1.7 ha per capita of productive land and water ecosystems. Since we are consuming more than can be replenished, pollution keeps building and resources keep dwindling. Climate change is only one of the outcomes of humans living beyond our means.

When the Stockholm Resilience Centre developed a framework to measure and track different areas of the planetary environment that humans were impacting, of the nine categories identified, climate change only ranked fourth.[24] It was ranked above the safe zone, into the "zone of uncertainty". In other words, it had passed the safe operating point. It could be nearing a

---

[23]Rees, W.. End game: the economy as eco-catastrophe and what needs to change. *real-world economics review, issue no. 87*, 2019

[24]The nine planetary boundaries, Stockholm Resilience Centre. Available at: www.stockhol mresilience.org/research/planetary-boundaries/the-nine-planetary-boundaries.html

threshold or tipping point, where there is a potential for a more considerable, possibly catastrophic change in the natural Earth System itself. Land system change is another area where we are operating in the zone of uncertainty. Two of the earth's systems are past the "zone of uncertainty" and into the "danger zone": Loss of biosphere integrity (biodiversity loss and extinctions) and nitrogen and phosphorus flows to the biosphere and oceans.

As we are starting to discover the long-term consequences that COVID 19 will have on the global economy and society as a whole, we can begin to see how fragile the world is. COVID 19 was relatively mild in terms of its fatality rate. One can only imagine what the impact of a deadlier virus spreading around the world would have been. With rising population levels, the risk of the emergence of infectious diseases will only continue to increase. As the climate warms, existing diseases that are more prevalent in the south are making their way to the northern hemisphere and older strains of common illnesses such as smallpox and influenza may also be released from the thawing permafrost.[25] As science evolves, as with most things, the capabilities of manufacturing bioweapons are growing. Even a low-cost bioterrorist attack could provoke global political tensions with an immediate and long-lasting fear cost at the social level.[26]

Maybe you are not much of an environmentalist and don't believe that you will be directly impacted by the way we treat our planet. However, if you look at the rise and fall of past civilizations, some signs indicate that we should be worrying.[27] In less than 5 years, the great Roman Empire lost over half of its land area. The reasons for its failure were many, including over-expansion, climatic change, environmental degradation and poor leadership. Research into past civilizations shows that many of the same issues can precipitate a collapse: climate change, ecological degradation, inequality and oligarchy, complexity, external shocks and bad luck. As a whole, humanity isn't doing well with many of these issues.

---

[25] Hannah, C., Melting Permafrost Could Mean Return of Ancient Diseases, The Science Times, Aug. 17, 2020. Available at: www.sciencetimes.com/articles/26903/20200817/melting-permafrost-mean-return-ancient-diseases.htm

[26] Tournier, J.-N., et al., The threat of bioterrorism, *The Lancet* **19**, 18-19, Jan. 1, 2019

[27] Kemp, L., Studying the demise of historic civilisations can tell us how much risk we face today, says collapse expert Luke Kemp. Worryingly, the signs are worsening, *BBC Future*, Feb. 18, 2019, Available at: www.bbc.com/future/article/20190218-are-we-on-the-road-to-civilisation-collapse, (Accessed on: Dec. 31, 2021).

The objective of this chapter was not to make you lose hope in our future. It was to make you question our cultural myth that humanity is progressing and evolving and that the future will be better than the past. Accepting that our future might not be as rosy as we believed is hard. As the next chapter explains, we need to mourn for the loss of the future that will not be. Once you have grieved for the loss of the future that you had imagined, you can start to plan and prepare for an uncertain future.

3

Mourn for our Lost Future

"The deeper that sorrow carves into your being, the more joy you can contain."

– Khalil Gibran

Abstract

For many, it can be hard to process that our future is fraught with uncertainty. In order to get through the anxiety this induces, we must mourn for the future we had imagined. This chapter explains how this mourning process works. It introduces *Finding your purpose*, which not only helps you get through the mourning process but helps improve your personal resilience at the same time.

Key Words: climate grief, eco-anxiety, mourning, purpose, anxiety.

The last chapter's description of the threats posed by climate change can be scary and depressing. It is also a growing cause of anxiety. From a survey of 10,000 people aged between 16 and 25, three-quarters of them felt that the future was scary and over half (56%) think humanity is doomed.[1] Given the rise of anxiety related to climate change and the environment, eco-anxiety was officially recognized by the American Psychological Association in

---

[1]Harrabin, R., Climate change: Young people very worried – survey, BBC News, Sep. 14, 2021. Available at: www.bbc.com/news/world-58549373, (Accessed on: Dec. 31, 2021)

2017 as a chronic fear of environmental doom.[2] For some, the condition is debilitating, leaving them feeling emotionally exhausted, depressed, lonely, lethargic and feeling helpless.

Many of the feelings brought about by this eco-anxiety are similar to the feelings we get when we lose someone close. It shouldn't come as a surprise that there is a growing body of evidence that more and more people are experiencing "climate grief" around the loss and anxiety related to the overall effects of climate change and its impact on our future.[3] When we don't recognize it as grief, it shows up as anxiety and depression. If you are genuinely going to accept that you need to start planning and preparing for a turbulent future, you first need to grieve for the imagined future that you had.

Grief is something that impacts us all in different ways. Two people going through the same loss will go through grief in their way at their own pace. I remember three specific events in my life where I was surprised by grief. Once for thinking I wasn't feeling sad enough and twice for feeling a level of grief that I had not expected.

The first experience was when I was a young child. One of my great-aunts passed away. It was someone I would see a couple of times a year. The only thing I remember of her is her house, which was filled with cuckoo clocks and whose living room looked exactly like a museum. The fact that the house left more of an impression on me than my aunt reflects my impressionable young age.

Her death is one of the few things I remember from when I was that young. I was in my bedroom and I was trying to force myself to cry. I was sitting there thinking that I wasn't normal, that I wasn't feeling sad enough. It's not something that I recall worrying about before or after that experience. It was just my first real experience facing death, and it wasn't what I had expected.

The next experience with grief that surprised me was in my early twenties. I was recently married and we were expecting our first child. Then after

---

[2]Comeau, N., 'Eco-anxiety' is a crushing weight for many young Canadians. And they say schools aren't doing enough, The Toronto Star, Feb. 23, 2020

[3]Pihkala, P., Climate grief: How we mourn a changing planet, BBC Future, Apr. 2, 2020, Available at: www.bbc.com/future/article/20200402-climate-grief-mourning-loss-due-to-climate-change

12 weeks of pregnancy, my wife had a miscarriage. What surprised me this time was how intensely my wife and I grieved for this lost child. It wasn't a child that we had come to know, yet we were both devastated. It gave me a whole new appreciation for all of the women I knew that had gone through miscarriages. It is sad news for friends and family to hear about a miscarriage, but when it happens to you, it crushes all of the plans and dreams that you had attached to this growing fetus.

First grieving too little for an extended family member's death and then grieving too much for the death of someone I had never known. Finally, my third surprising experience with grief wasn't associated with death at all but with the birth of my firstborn son. Cedric was born one week early at a midwifery birthing house in Montreal and seemed like a small healthy baby. However, once the midwife realized he was less than 5 lbs, the standard procedure required her to take his blood sugar reading. She tried twice, and it seemed that the machine was broken since the readings were abnormally low. Nevertheless, they called the ambulance to get him checked out at the hospital.

When the paramedic got to the birthing house, it became apparent that there was a problem. The baby was getting weak and my wife and I don't know for sure, but it seemed that he even stopped breathing for a bit because the paramedic was busy trying to get him oxygen and she seemed quite worried. He ended up at the neonatal intensive care unit and it took three weeks before his pancreas started working to regulate his blood sugar. On his second day at the hospital, the cardiologist found a hole in his heart. At three months, he had open-heart surgery, and they implanted a Gortex patch to repair the hole. It's not a sad ending; now, 15 years later, he's a healthy young teenager. So, where did the grief come from?

It turns out that when your baby is born and is not the perfectly healthy baby that you were expecting, it is normal to experience grief. You have to mourn for the baby and experience that you had imagined in your head. It didn't diminish the amount of love we felt for our son. The grief was not for him. It was for our imagined experience and our imagined child.

After these three personal experiences, I have a better understanding of grief. It wasn't a surprise when I read an article from the Harvard Business Review explaining that what many people were going through emotionally with the COVID 19 pandemic and the associated lockdown was grief.[4] According to David Kessler, a grief expert, we were collectively grieving for the loss of normalcy, the fear of the economic toll, the loss of connection. He explains that there are different types of grief and we are feeling anticipatory grief. In the article, he explains that anticipatory grief is the feeling we get about what the future holds when we're uncertain. Although it is often associated with death, it can be related to a bad health diagnosis and it can also be broader. We can feel it when our future suddenly becomes uncertain when we feel a storm coming.

When it comes to dealing with grief, many of us have heard about the five stages of grief:

Denial– The first reaction is denial, which is where most of the world seems to be concerning our future with climate change. People tend to cling to a false, preferable reality.

Anger – When denial cannot continue, people often move on to anger. They may look for someone or something to blame. *It's the oil companies' fault; they were purposefully raising doubt around climate change science even if they knew it was real.*

Bargaining – The third stage involves the hope that you can avoid a cause of grief. This could manifest itself in someone vowing to reduce their climate impact to lessen the likelihood that climate change will happen.

Depression – Depression is when you can become overwhelmed with feelings and feel futility. *What's the point of doing anything to prevent or adapt to climate change? There is nothing I can do to stop climate change.*

Acceptance – In the last stage, people embrace the inevitable future. *I can't fight it; I may as well prepare for it.*

According to Kessler, understanding the five stages of grief is a start. But everyone experiences grief differently and it's not a linear process. You

---

[4]Derinato, S., That Discomfort You're Feeling Is Grief, Harvard Business Review, Mar. 23, 2020. Available at: https://hbr.org/2020/03/that-discomfort-youre-feeling-is-grief, (Accessed on: Dec. 31, 2021)

might jump from one stage to another and back again. Acceptance is the key. We find control in acceptance. Before we can accept it, we need to identify it and name it. *"There is something powerful about naming this as grief. It helps us feel what's inside of us. When you name it, you feel it, and it moves through you. Emotions need motion"*. Once you put a name to it, you can find resources to help you through it. For example, the BBC Future's Climate Emotions series aims to help you go from fear and anxiety to hope and healing by helping you understand all of the different emotions associated with climate change.

You might find yourself at one or a number of these stages of grief. It could be that you accept it one day and then go back to being angry or depressed. After going through your grief, you should find yourself in acceptance. If you are genuinely going to plan and prepare for an uncertain future with drastic impacts related to climate change, you need to accept and embrace this future.

3.1 Finding Your Purpose

> *"He who has a why to live can bear almost any how."*
> – Nietzsche

David Kessler recently revisited his five stages of grief, adding a sixth stage, finding meaning.[5] Although his book is more focused on finding meaning after losing a loved one, finding meaning is just as relevant when grieving that climate change is real and will worsen in the years to come. Helping the world get climate change under control or simply helping the world manage the impacts of a changing climate can give meaning and purpose to your life.

Purpose, defined as a commitment to make a meaningful contribution to the world, gives direction to life and is associated with physical, social and psychological wellbeing. Finding meaning or purpose in your life is an important element of resilience. Whether you are a parent that is focused on raising a happy, healthy child, a volunteer helping others in the community, etc., your purpose gives you the will to persevere. It is the beacon that can

---

[5]Kessler, D., Finding Meaning: The Sixth Stage of Grief, Sep. 1, 2020. Accessed at: https://grief.com/sixth-stage-of-grief/

lead you out of the darkness. In some cases, someone's existing purpose helps them through a trying time in their life. Whereas in many cases, people find meaning in response to adverse events, such as pursuing a medical research career after a beloved aunt died of cancer or mentoring younger teens in a community youth group after getting caught up in drugs and violence.[6]

Like many of the other elements of resilience, purpose in life improves both health and longevity.[7] There are so many problems in the world and many of them will only get worse with climate change. From rising inequality to human rights abuses and world hunger, if most people are not actively working towards making a meaningful contribution to the world, things will only get worse. There is only so much you can do as an individual, but if we are all working towards a greater good, we can see real change.

I have personally tried to couple my purpose with my career, in the sense that when looking for a job, I look for something that will help me work towards addressing causes that I am passionate about (i.e. climate change). When I start to feel that my job is not having a meaningful contribution, I tend to start looking for another job. Similarly, when work is hard and stressful, I look towards the contribution that I am making to give me the stamina to keep going.

Even though I like the idea of finding a career that aligns with my purpose, I know that is not for everyone. You might take on a job because you enjoy doing it, even though it doesn't contribute to the greater good. Or it might be that you are doing a job to provide for yourself and your family. You can achieve your purpose outside of your job. Whether it is contributions you make to your family, the community or society at large, there are other ways of making a difference. The key is that you feel that you are making a difference in something that you find important.

You can teach purpose. In her book "Teaching for Purpose", Heather Manlin explains that a growing number of educators see developing students'

---

[6]Malin, H., Teaching Purpose for Resilience and Flourishing, The Blog of Harvard Education Publishing, Jan. 31, 2019. Available at: www.hepg.org/blog/teaching-purpose-for-resilience-and-flourishing

[7]Schaefer, S., et al., Purpose in Life Predicts Better Emotional Recovery from Negative Stimuli, PLoS One **8** (11), 2013.

purpose as not only possible but essential.[8] By exploring and reflecting on their core values, considering their role in society and what they can contribute to their community, students can develop a sense of purpose. Although finding your life's purpose as a teenager would be ideal, it is never too late to do it. If you feel like your life lacks meaning or purpose, you should embark on a journey of self-discovery to find how you can contribute to making the world a better place, as corny as that may sound.

---

[8]Malin, H., Teaching Purpose for Resilience and Flourishing, The Blog of Harvard Education Publishing, Jan. 31, 2019. Available at: www.hepg.org/blog/teaching-purpose-for-resilience-and-flourishing

4

Plan for Personal and Family Resilience

"As Darwin himself was at pains to point out, natural selection is all about differential survival within species, not between them."

– Richard Dawkins

Abstract

Chapter 4 introduces the resilience plan outlined in the remainder of the book and the five areas of resilience that it covers. It starts by explaining what resilience is and why it's important. It presents the types of things that should be considered as we plan for the future. Finally, it stresses the fact that even though this book focuses on resilience, people shouldn't give up trying to reduce greenhouse gas emissions and limit the impact of climate change.

Key Words: resilience, fitness and health, community building, learning and education, financial, physical preparation

According to Google Trends, Google searches for the term "Resilience" nearly doubled between December 2019 and April 2020. After the pandemic has hit or after other trouble finds you, it can be too late to make yourself more resilient. The term resilience is starting to get over-used. It is like the term sustainability or even the color "green". It has been associated with so many different things, many of them for marketing purposes, that the word starts to lose its meaning. While searching for a definition of resilience and for guidance on how to be more resilient, I found quite a few sources and there was quite a range between them. Some are centered around business resilience, whereas others focused primarily on emotional or psychological resilience. One definition that I found that I liked was from Dorie Clark, the author of "Reinventing You", who said:[1]

> *"Resilience is a combination of being able to pick yourself up when there are setbacks, but also it is about having the kind of cross-training necessary to be flexible in an uncertain world where we don't know what is around the corner."*

The key is that we don't know what is around the corner. We know that some dark clouds are on the horizon and what they bring is anyone's guess. An example of how becoming more resilient can help you face different challenges is looking at my in-law's beef and seed-potato farm in Pemberton. In the early 2000s, a series of storms hit at once. Rising oil prices meant that fuel costs soared, increasing operating costs. Rising oil prices helped increase the value of the Canadian dollar, which hurt the farm since most of their potato sales were to US customers in US dollars. Finally, a few cows in Alberta were found to have "mad-cow disease", effectively closing the border to sales of beef to the US and the world, severely impacting the price of all cattle sold in Canada. As a result, my in-laws needed to ask their three adult children to send money home every month to make ends meet.

At the same time, my in-laws tried to diversify the farm to not solely depend on seed potatoes and cattle income. They tried many things that didn't work as well as they hoped. But in the end, the farm started selling potatoes to the local restaurants and my mother-in-law got a job managing the local feed store, which she ended up buying. The farm now also grows and sells carrots with potatoes. So now, when COVID 19 hit, when the commercial restaurant potatoes stopped selling, they could shift some of the sales to their seed market, and they still had a few revenue streams to keep the farm afloat. Having multiple sources of revenue means that they can weather different storms. And with farming, climate change is bringing with it more severe storms that will test farmers' resilience.

---

[1]Hannon, K., To Build Emotional Strength, Expand Your Brain, The New York Times, Sep. 8, 2020. Available at: www.nytimes.com/2020/09/02/health/resilience-learning-building-ski lls.html (Accessed on: Dec. 31, 2021)

4.1 Plan for What?

Preppers are people or groups of people preparing for the fall of civilization, which they think is imminent. They have bunkers with lots of food, generators with fuel, weapons and ammunition, etc. They believe their preparation will allow them to survive as the society around them crumbles. It could be that society will collapse in the next year or two and that these preppers will be ahead of the curve. On the other hand, society might be in for a slow decline where there is never really a call for all of their preparation. No one knows what exactly will happen and when it will happen. So you need to make yourself more resilient to any eventuality, or another way of looking at it is that you should make yourself less vulnerable when disaster does hit. If you have a good-paying job in a coal mine in a mining town, now is not the time to buy a big new house with a large mortgage. It is only a matter of time before coal mines will need to be shuttered as the world moves towards cleaner fuels. Now would be the time to pay off your debts and maybe learn a new skill that could be transferrable if you lose your job.

There are many ways that you can be impacted by climate change, either directly or indirectly. A direct impact would be that your house gets flooded by the 100-year storms that seem to be happening every couple of years. A tornado that blows through your town when it didn't use to happen and rips your roof off. A forest fire goes through town and destroys your house or even your whole town as it happened in Lytton, British Columbia, in 2021 after breaking the record for the hottest temperature ever recorded in Canada. Somewhere in the world, people are directly impacted by climate change every day. As the frequency and severity of these storms keep increasing, we will likely no longer be able to rely on the insurance industry to help us out.

There are already large areas with hundreds of thousands of homes that can no longer get flood insurance. Homes near forests might see their fire insurance premiums become unaffordable. At our farm, the fire insurance would be over $10,000 per year, something that the farm could no longer afford. The government will come in and bail out communities that are no longer covered by insurance, but this is generally limited in the overall amount that they give. If it's flood damage, the payouts generally only happen once. Eventually, there is a high likelihood that the insurance industry and the government will no longer cover the losses.

Climate change can also impact you indirectly. With the increasing severity of fires and droughts hitting California, their farms will no longer be able to continue to supply grocery stores across North America with fresh produce all year round. We may need to get used to having more expensive food with less variety, especially in winter. Other indirect impacts could be a recession, or it could be that your industry gets disproportionally impacted. In British Columbia, climate change allowed the mountain pine beetle to kill countless trees, which led to many layoffs in the forestry sector. As climate change gets worse and worse, you can imagine that eventually, global leaders will be forced to implement policies that will have a meaningful impact. Once this happens, the fossil fuel industry will be severely impacted.

4.2 What's the Plan? Become More Resilient

The plan that I outline in this book will help you become more resilient. The good thing is that this plan is not limited to only climate change impacts. It will help prepare you for the unexpected. In addition, most of the plan elements can help improve your day to day life, so it won't be a waste of time if luck would have it and nothing bad ever happens to you. The plan is broken down into five areas in your life where you can start implementing changes today.

Fitness and Health: The first part of the plan focuses on physical and mental health.

Learning and Education: The next part focuses on skills and knowledge that you and your family can learn to become more resilient.

Community Building: Planning around developing a supportive community where you can rely on each other in times of need.

Financial: Steps you can take to make your financial health more robust to unexpected expenses or disruptions of income.

Physical Preparation: Finally, what tools you can buy and renovations that you can do to your house that will make you better prepared to face the future.

Figure 4.1 Five different categories of the personal resilience plan.

These different areas are further broken down in the book, as shown in the
Figure 4.2, to help you develop a more thorough and complete plan. In order
to get started on your journey to resilience, you can fill in your Personal
Resilience Worksheet, which you can either download electronically at
www.remicharron.ca/book or use the sheets provided in the
Appendix.

4.3 Don't Forget Mitigation Efforts

Although I believe it is time that we admit to ourselves that we will not
achieve our 1.5 °C or 2 °C global warming target and that we should start
planning for the potential repercussions, it is not time to stop trying to

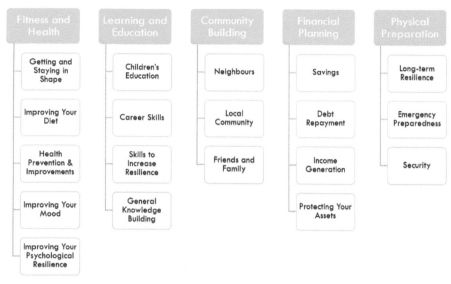

Figure 4.2   Sub-categories of the personal resilience plan.

achieve significant reductions in global GHG emissions. After all, an ounce of prevention is worth a pound of cure. The more that we can do to limit our global GHG emissions, the less severe the outcomes will be and the more likely it is that you will be successful in your personal resilience plan. As individuals, the most meaningful impact we can have is to elect the right politicians and demand more serious and meaningful change.

You should do your part and reduce your GHG emissions. However, if you're like me, it can be disheartening to put a lot of effort into reducing your personal GHG emissions when the rest of the world is not committed to achieving our targets. The good news with personal resilience is that if you do make efforts to make yourself more resilient, these efforts will make a difference in your life.

5

Physical Health

"Every day is another chance for you to get stronger, to eat better, to live healthier and to be the best version of you."

– Unknown origin

Abstract

The first area of resilience that is covered in the book is Physical and Mental Health. This area of the plan is broken down into two chapters, starting with a chapter on physical health. The chapter breaks down physical health into three categories: Getting and staying in shape, Improving your diet and Health prevention and improvements.

Key Words: fitness, health, diet, exercise, prevention.

Fitness and health, both physical and mental, are key components to resilience. Research has demonstrated a strong link between mental and physical health.[1] These act together as a positive feedback mechanism. Better physical and mental health tends to lead to more physical activity, which then leads to even better mental and physical health. It can also act as negative feedback with depression leading to less physical activity, which then further deteriorates your physical and mental health. The fitness and

---

[1]Ohrnberger, J., Fichera, E., Sutton, M., The relationship between physical and mental health: A mediation analysis, *Social Science & Medicine* **195**, 42-9, 2017

health section of the resilience plan is divided into two chapters, starting with physical health.

In terms of resilience, you just never know when being in shape will help you. At 17 years old, Juliane Koepcke flew over the Amazon rainforest on Christmas eve in 1971, when the plane crashed after it was struck by lightning, Juliane was its sole survivor.[2] Despite a broken collarbone, deep cuts on her legs and a ruptured ligament in one knee, Juliane waded down a stream through knee-high water for 10 days before being rescued. Her only source of food was a bag of candy she found at the crash site. She relied on her physical and mental strength and the knowledge she gained in the jungle, where she was homeschooled by her parents when she was 14.

Although few of us will be tested to that extent, it is not hard to find examples of people who have become so out of shape that it impacts their daily lives, let alone when the unexpected occurs. A man I know is a good 100 lbs overweight with Type 2 diabetes before COVID 19 hit. With his pre-existing conditions and the fact that obesity is one of the highest risk factors associated with severe illness or death from COVID-19, he decided to self-isolate for the duration. Staying at home without seeing any friends for months on end, with a frightening global pandemic, was hard emotionally. Unfortunately, that led to more eating, less exercise and more weight. Now the extra weight is even more of a problem, and he can't move around without a mobility device.

You never know when being in shape can help save you or when being out of shape can hurt you. The good thing is that the list of benefits for good physical health is long and keeps growing as more studies are done. Physical activity benefits start right after a session of moderate-to-vigorous physical activity by improving thinking for children 6 to 13 years of age and reduced short-term feelings of anxiety for adults.[3] Among other benefits, regular physical activity has been found to:

1. Help keep your thinking, learning, and judgment skills sharp as you age.
2. Reduce your risk of depression and anxiety.

---

[2]Koepcke, J., How I survived a plane crash, BBC News, Mar. 24, 2012, Available at: www.bbc.com/news/magazine-17476615 (Accessed on: Dec. 31, 2021)

[3]Benefits of Physical Activity, Centers for Disease Control and Prevention, Available at: www.cdc.gov/physicalactivity/basics/pa-health/index.htm (Accessed on: Dec. 31, 2021)

3. Help you sleep better.
4. Boost your overall energy levels.
5. Enhance your sex life.
6. Reduce health risks from cardiovascular disease, diabetes, some cancers, etc.
7. Strengthens your bones and muscles, helping reduce falls and fractures.
8. Help prevent excess weight and maintain weight loss.

According to the US Centre for Disease Control, just 150 minutes a week lowers all-cause mortality by 33%. The more research is done, the more benefits of exercise are found. One new study done on lab mice found that regular exercise helped lab mice remain psychologically resilient even when their lives seemed filled with threats.[4]

In addition to psychological health and wellbeing that will be discussed separately in the next chapter, I have broken down physical fitness and health into three sub-categories:

1. Getting/staying in shape
2. Improving your diet
3. Health prevention and improvements.

I will provide more details on each of these and some example activities that you could include in a resilience plan.

5.1 Getting and Staying in Shape

> *"Exercise is the key not only to physical health but to peace of mind."*
>
> – Nelson Mandela

As indicated, there are countless benefits for exercising to either get or stay in shape, which will all play a part in making you more resilient. It used to be that most of the jobs available relied on manual labor. There are still jobs

---

[4]Reynolds, G., Exercise May Make It Easier To Bounce Back From Stress, The New York Times, Sep. 9, 2020, Available at: https://www.nytimes.com/2020/09/09/well/move/Exercise-stress-resilience.html (Accessed on: Dec. 31, 2021)

in the construction and farming industry where physical strength and stamina would be a critical asset to both get and succeed in the job. These jobs could be potential fallback options if your particular industry faces cutbacks and layoffs. There can also be other benefits such as meeting new friends and growing your community, which is another area of the resilience plan discussed later in this book.

Example activities that could be part of your plan:

1. Setting a distance goal for your preferred activity, be it walking, running, biking, swimming, rollerblading, etc.
2. Joining and participating in a sports team.
3. Sign up and go to the gym or exercise classes.
4. Joining and participating in a local cross-fit gym/club.
5. Registering and then training for a race.

Activities that you choose should both align with your preferences as well as your current level of fitness. There is no use in setting a short-term goal that is too hard leading to injury or discouragement. It's okay to bribe yourself or use measures that will encourage you to follow through. I had an apartment in Vancouver where I spent my weekdays, and it did not have a TV. My local gym had elliptical machines that were connected to cable TV. It not only encouraged me to go regularly, I generally extended my workouts so that I could finish the show I was watching.

Knowing myself, there are other tricks I use to keep exercising. I started running regularly but soon found that I was missing more and more days. I decided to register for a half-marathon to give me a goal and the needed motivation to keep running. As an added bonus, I convinced my wife to register as well. This not only gave me an occasional training partner, but the shared goal was also enough to keep me running on a regular basis. Similarly, when I join team activities, the obligation I feel to my team never fails to motivate me to go to a workout or game. I am also cheap by nature. When I paid $100/month to join my local gym, I hated to think about how expensive it was getting per workout if I missed a few days at the gym.

Knowing yourself, pick activities that will keep you motivated to continue over the long term. Once you think of goals that interest you, they can be included in your Personal Resilience Worksheet, which you can either download electronically at www.remicharron.ca/book or use the sheets provided in the Appendix. The following is an example of a plan for getting and staying in shape.

Getting and staying in shape	
Start this year	Reach my goal of 10,000 steps/day for 6/7 days in the week.
Start next year	Train for and participate in the local half-marathon.
Medium term, years 3–10	Run at least every other day for a minimum of one hour. Travel to a new half-marathon every year as a "vacation." Improve my time most years.
Long-term, over 10 years	Workout at the gym at least 3 times a week for an hour Run at least 20 km a week to keep my endurance.

5.2 Improving Your Diet

"Tell me what you eat, and I will tell you what you are."

– Anthelme Brillat-Savarin

The benefits of eating healthy are similar to those described for regular exercise, including:[5]

1. Weight loss.
2. Reduced cancer risk.
3. Diabetes management.
4. Heart health and stroke prevention.
5. Establishing healthy habits for the next generation.
6. Strong bones and teeth.
7. Better mood.
8. Improved memory.
9. Improved gut health.
10. Good night's sleep.

Again, being healthier and enjoying all of these benefits will help make you more resilient to face any emerging challenge. When the US Air Force

---

[5]Crichton-Stuart, C., What are the benefits of eating healthy?, Medical News Today, Available at: www.medicalnewstoday.com/articles/322268, (Accessed on: Dec. 31, 2021)

researched how they could make their members and their families more resilient, they found that nutrition contributes to resilience.[6] It helps service members maintain a healthy weight, protecting them against diet-related diseases that affect physical and cognitive functions and reducing their vulnerability to stress and depression.

Over the years, research related to nutrition continues to evolve and change. Depending on what study you look at, certain foods can be harmful, good or neutral, many of them contradictory. This can include fats, grains, salt, gluten, dairy, coffee, red meat, etc. It can become confusing to the point of people tuning out what the latest and greatest research has to say. Despite all of this, there are definite ways of improving your diet.

Cook for yourself: Unless you go to a health-food restaurant or you limit yourself to the 1-2 healthier options on the menu, restaurant food is often high in calories, fat and salt and is generally not as nutritious as a home-cooked meal.

Eat less deep-fried foods: Limit the number of fried foods that you eat. Fried foods are significantly higher in fat and calories than their non-fried counterparts.[7] Restaurants often use processed vegetables or seed oils, which can form trans fats when heated. The more fried foods you consume, the higher your risk of developing type 2 diabetes, heart disease and obesity.

Eat less processed food: Almost all of the processed food made in factories is bad for you as they tend to be high in sugar, artificial ingredients, refined carbohydrates and trans fats, while at the same time being low in nutrients and fiber. Unfortunately, these processed foods are often fast and convenient, taste relatively good and are usually cheaper than the healthier alternative. Food manufacturers use unhealthy levels of added sugar, sodium and fat that often trigger addiction responses.[8] Cutting back on the amount of processed food you eat will help improve your overall health.

---

[6]Florez, K., Shih, R., Martin, M., Nutritional Fitness and Resilience - A Review of Relevant Constructs, Measures, and Links to Well-Being. RAND Corporation, 2014.

[7]McDonell, K., Why Are Fried Foods Bad For You? Healthline, Nov. 19, 2017. Available at: www.healthline.com/nutrition/why-fried-foods-are-bad, (Accessed on: Dec. 31, 2021)

[8]Moawad, H., Underlying Mechanisms of Highly Processed Food Addiction, Psychiatric Times, Oct. 28, 2019. Available at: www.psychiatrictimes.com/view/underlying-mechanisms-highly-processed-food-addiction

Eat more fruits and vegetables: As a general rule, fruits and vegetables should take up at least half of your plate, with a heavier emphasis on vegetables as they are generally rich in fiber, vitamins and minerals, while also being low in sugar, sodium and fat.

Cut out refined and added sugar: Added sugar introduces calories with no added nutrients and has been linked to weight gain and various diseases like obesity, type 2 diabetes and heart disease. The average American's daily intake of added sugar is 77 grams per day, which equals 19 teaspoons or 306 calories.[9] The maximum recommended intake for men is 37.5 grams or nine teaspoons and 25 grams or six teaspoons for women. One regular can of pop contains roughly 40 grams of sugar or more than what anyone should consume in a day.

Control when and how much you eat: It's a lot easier to limit how many calories you eat in a day when you eat less at restaurants, cut back on processed foods, eat more fruits and vegetables and limit the amount of added sugar you consume. Even if you eat extra calories of broccoli, the odds are it won't be detrimental to your health. However, if you are exercising and eating healthy yet are still gaining weight, it could be that you are eating too much. It could also be a factor in when you are eating. Studies show that when food is consumed late at night, the body is more likely to store those calories as fat and gain weight rather than burn it as energy.[10]

Drink less alcohol: The good news if you like to indulge in a glass of wine or an occasional beer is that moderate alcohol use for healthy adults (1/day for women, 2/day for men) has been linked to some minor potential benefits (reducing heart disease and stroke).[11] However, if you drink more than that, the benefits disappear, and the health risks quickly begin to pile up (more cancers, pancreatitis, dementia, sudden death, heart muscle damage, stroke, high blood pressure, liver disease, suicide, serious injury or death).

---

[9]Gunnars, K., Daily Intake of Sugar – How Much Sugar Should You Eat Per Day? Healthline, Jun. 10, 2021. Available at: www.healthline.com/nutrition/how-much-sugar-per-day, (Accessed on: Dec. 31, 2021)

[10]Van Allen, J., Why eating late at night may be particularly bad for you and your diet, The Washington Post, Aug. 24, 2015

[11]Mayo Clinic Staff, Alcohol use: Weighing risks and benefits, Mayo Clinic, Dec. 11, 20121. Available at: www.mayoclinic.org/healthy-lifestyle/nutrition-and-healthy-eating/in-depth/alcohol/art-20044551

The idea of improving your nutrition is not about going on a diet but gradually transitioning your current eating habits to healthier choices. The following are examples of some of the successful ways that I have modified my diet over the years:

- Limited going out to the restaurant to once per week.
- Over a year, I gradually reduced the amount of sugar that I added to my coffee until I got used to drinking unsweetened coffee.
- When baking, I cut back on added sugar by at least 50% and, in some recipes, I replaced sugar entirely by adding bananas or unsweetened apple sauce.
- At the restaurant, I chose only one among appetizers, drinks or desserts.
- I replaced soft drinks with carbonated water flavored with a few natural fruit juices.
- I only put two cans of beer in the fridge at a time to limit my daily intake to a maximum of two.
- Avoid eating after supper.
- Signed up with a local farmer to receive a weekly box of locally grown vegetables. Supporting local farmers also helps make your community more resilient in the long run.

Once you think of action items and goals that you would be interested in following, you can include them in your Personal Resilience Worksheet, which you can either download electronically at www.remicharron.ca/book or use the sheets provided in the Appendix. The following is an example of a plan for improving your diet.

Improving your diet	
Start this year	Reduce restaurant outings to one supper/week and one lunch/week. Reduce intake of pop to one per week. Stop buying junk food when I stop at the gas station
Start next year	Above items plus: Cook vegetarian meals three nights/week Limit alcohol intake to a one-liter bottle of wine per week.
Medium term, years 3–10	Above items plus: Sign up with a local farmer to receive a weekly box of locally grown vegetables. Eat all of the vegetables in the bin each week. Stop drinking all sugary beverages. Introduce more fermented foods into my diet.
Long-term, over 10 years	Maintain the healthy habits developed to date. Explore preserving my own food.

5.3 Health Prevention and Improvements

"Another flaw in the human character is that everybody wants to build and nobody wants to do maintenance."
– Kurt Vonnegut

In terms of fitness and health, we talked about the importance of exercising and diet, but this third category is more about keeping your body in top shape. This might not apply to everyone right away, but it concerns addressing health issues before they become a major problem. You could include getting vaccinations, both regular flu shots and those for specific diseases (e.g. shingles, hepatitis, etc.). If you have a mole or lump that concerns you, you should go to the doctor to get it checked out.

You could also include proactively doing something to make your body more resilient. Some people have high cholesterol and then start taking cholesterol-lowering medication for the rest of their lives. Simple dietary changes have been shown to help manage cholesterol.[12] Being in control of managing your heart health might be more resilient than relying on pills from the pharmacy. Diet and exercise can also be used to help manage type 2 diabetes. Some people can manage their blood sugar using a combination of exercise, diet and weight loss. Again, if you can manage your condition on your own, under the supervision of a doctor, it can be more resilient than requiring pharmaceuticals for the rest of your life.

Other potential items to include here would be getting procedures that could improve your physical health. For example, it is possible that getting laser surgery on your eyes can be more resilient as you may no longer be reliant on glasses to function. Alternatively, if you have an old knee or another joint injury preventing you from doing some activities, now might be the time to get physiotherapy or other treatments to finally heal properly.

Once you think of action items and goals that you would be interested in doing, you can include them in your Personal Resilience Worksheet, which you can either download electronically at www.remicharron.ca/book or use

---

[12]How to lower your cholesterol without drugs. Harvard Health Publishing, Apr. 15, 2020. Available at: www.health.harvard.edu/heart-health/how-to-lower-your-cholesterol-without-drugs

the sheets provided in the Appendix. The following is an example of a plan for health prevention and improvements.

Health prevention and improvements	
Start this year	Book an appointment with the doctor for a flu shot and discuss other vaccinations that may be available. Get a physical and blood test done to get a baseline on my current health. Examine potential deficiencies (iron, vitamins, etc.), blood pressure, cholesterol levels and blood sugar—research and implement diet changes that would address issues, if any. Book a long-overdue appointment with the dentist for a cleaning and checkup. Book the next one to get on a regular schedule.
Start next year	Start physiotherapy to address shoulder pains that started when competitive swimming as a teenager. Get an eye exam and book the next one to get on a regular schedule.
Medium term, years 3–10	Continue regular checkups at doctor, dentist and eye doctor.
Long-term, over 10 years	Start regular colonoscopies and PSA level testing to check for colon and prostate cancer.

No one can predict the adversity they will face in life. Whether it's an accident that leads you to a life or death situation where health will be an asset or you want to build a more robust immune system to be ready for the next pandemic, now is the time to start. Improving your overall health and fitness will be a good foundation that will help improve all aspects of your life, including helping you become more resilient. On the same level as physical health, emotional well-being is interlinked and just as important. The next chapter continues with the fitness and health theme but focuses on its psychological aspects.

6

Mental Health

"Resilience is very different than being numb. Resilience means you experience, you feel, you fail, you hurt. You fall. But, you keep going."

– Yasmin Mogahed

Abstract

The Mental Health chapter is broken down into two categories: Improving your mood and Improving your psychological resilience. Many personal resilience books focus primarily on helping you build your psychological resilience. This factor can help explain why two people experiencing the same trauma can react in very different ways, from being the spark to transform their life for the better, to using it as a perpetual excuse why their life is heading down the drain.

Key Words: anxiety, depression, psychological resilience, gratitude, nature therapy, emotional awareness.

Psychological resilience is a big emerging field. Quite a few of the books that have been written to help build your personal resilience focus almost exclusively on the psychological/emotional realm. Two people facing the same situation (e.g. job loss) will react in very different ways. Some would let that event define them and essentially give up, while others take it as an opportunity to do something great. Essentially how you react emotionally to adversity will determine how resilient you are.

49

Nelson Mandela's story is a perfect example of someone facing great hardship that used emotional resilience and spirit to get out of it stronger. Mandela spent 27 years in harsh prison conditions for opposing South Africa's apartheid system.[1] Despite his imprisonment, Mandela continued to act as a leader and mobilized his fellow political prisoners. He endured harsh conditions as black prisoners ate more poorly than Indian/Asian or colored (people of mixed race) prisoners. Black men were forced to wear shorts and sandals, even in winter, while other prisoners could wear pants and shoes. Mandela and his fellow activists spent more than a decade of hard labor breaking rocks in a lime quarry. Contact with the outside world was limited to one letter and one 30-minute visit every six months.

On February 11, 1990, Mandela was released at 71 years old. It is what he did next that is remarkable. He left prison without holding a grudge against his captors. He said, *"As I walked out the door toward the gate that would lead to my freedom, I knew if I didn't leave my bitterness and hatred behind, I'd still be in prison."* He went on to win a Nobel Peace Prize in 1993 and to become South Africa's first democratically elected president in 1994.

On the opposite end of the psychological resilience spectrum would be people that commit suicide soon after something devastating happens. Impulsive suicide attempts can have as little as a few minutes between the thought of committing suicide and the attempt itself. Impulsive suicides are often related to interpersonal conflicts and suicide as a means of reducing tension or escape.[2] It could be that the devastating event is all it takes or it could be coupled with depression or other factors. When I was in my first year of university living in residence, a student committed suicide by jumping off the building. It turns out he had just been caught cheating on a test, and there would be a disciplinary hearing. Although I don't know the whole situation, I can only imagine that disappointing his parents was too much to bear. Now, as a professor, when I am forced to discipline students for cheating, I always worry about their ability to cope with the consequences of their actions.

---

[1]McRae, M., The story of Nelson Mandela - A will that could not be contained, Canadian Museum for Human Rights, Available at: https://humanrights.ca/story/the-story-of-nelson-mandela (Accessed on: Dec. 31, 2021)

[2]Kim, J., et al., Characteristic Risk Factors Associated with Planned versus Impulsive Suicide Attempters, Clinical *Psychopharmacology and Neuroscience* **13**(3): 308–315, 2015

If escape by drugs or alcohol (or at the extreme, suicide) is one of your potential reactions to severe adversity, you will have difficulty making it through our challenging future. COVID 19 led 13% of Americans to report starting or increasing substance use as a way of coping with stress or emotions.[3] Overdoses also spiked 18% in the US compared to pre-pandemic levels. Luckily, there are ways to help train yourself to become more psychologically resilient. I have split these into two different categories. The first, *improving your mood*, are activities and strategies you can regularly employ to make you feel happier. Although learning techniques to improve your mood will help make you more resilient, the second category, *improving your psychological resilience*, are activities geared specifically towards improving resilience.

6.1 Improving Your Mood

> *"If Laughter cannot solve your problems, it will definitely DISSOLVE your*
> *problems; so that you can think clearly what to do about them"*
> – Dr. Madan Kataria

Everybody strives to be happy or happier with their lives. If you are happy, is there anything else that you really need? If everyone is striving for happiness, then learning and practicing techniques and strategies that will make you happier should be a no-brainer. In this section, I list strategies that have been proven to be effective. These strategies can not only play a part in helping you be happier, they can also help replenish your tank when you are suffering from emotional fatigue.

Exercise

We saw in the last chapter that diet and exercise could help alleviate depression. For the last couple of years, I have started running and exercising more and treating depression is one of my primary motivators. I have suffered through mild to moderate depression that seems to ebb and flow for all of my adult life. For years I just kept it to myself. Then for some years, I was taking antidepressants. However, I didn't like the thought of

---

[3] Abramson, A., Substance use during the pandemic, *Monitor on Psychology* **52** (2), 22, 2021.

being dependent on drugs for years on end. It did not seem resilient and I don't trust that there wouldn't be long-term side effects. When I found more research showing that exercise was as or more effective than antidepressants and other treatment methods[4], I decided that I could rely on exercise rather than drugs to help. I found that it definitely has helped. Like antidepressants, it is not a magic solution that will rid me of depression, but it does help lift my mood and be less grumpy. I found it especially helpful during COVID 19. After being cooped up in my office all day, running is an excellent way to leave the worries of work behind.

Laughter and Tears

Both laughter and tears have been shown to be beneficial to help improve your mood. If you are feeling stressed or depressed, it can be difficult to laugh. If something makes you laugh out loud, from funny movies to looking at clips of epic fails, indulging in these can make you feel better. Benefits of laughter include reducing anxiety, decreasing stress hormones, lowering blood pressure, boosting your immunity, lowering depression, helping the respiratory and cardiovascular system.[5] The good news is that even if you don't feel like laughing, fake laughter can trick your mind and body, resulting in some of the same benefits. Laughter yoga and laughter clubs essentially have people come together regularly to laugh out loud to reap the benefits.

Fortunately, it is not just laughter that has health benefits. Shedding tears also has many benefits.[6] Crying is a natural way to regulate your emotions. It helps you calm yourself and reduce your distress. Shedding emotional tears releases oxytocin and endorphins, which can both help reduce pain and improve your mood. Tears contain several stress hormones and other chemicals, which may be a natural way for your body to get rid of these. If ever you have been feeling a build-up of stress and anger that have been

---

[4]Netz, Y., Is the Comparison between Exercise and Pharmacologic Treatment of Depression in the Clinical Practice Guideline of the American College of Physicians Evidence-Based? *Frontiers in Pharmacology* **8**, 257, 2017.

[5]Rosenfeld, J., 11 Scientific Benefits of Having a Laugh, MentalFloss, Apr. 11, 2018, Available at: www.mentalfloss.com/article/539632/scientific-benefits-having-laugh (Accessed on: Dec. 31, 2021)

[6]Burgess, L., Eight benefits of crying: Why it's good to shed a few tears, Medical News Today, Oct. 7, 2017. Available at: www.medicalnewstoday.com/articles/319631#benefits-of -crying (Accessed on: Dec. 31, 2021)

relieved by a good cry, you have felt the benefits of tears first-hand. You can even trigger these tears by reading a sad book or watching a sad movie. Even though you start the tears from outside sources, it can help you deal with the build-up of stress troubling you.

Journaling

Learning to be grateful for what you have instead of dwelling on what you don't have is an effective way to improve your happiness, whatever the circumstances may be. Taking a few minutes every night or a half-hour every week to write down in a gratitude journal what you are grateful for has been proven to be an effective strategy to boost happiness, lower stress, improve sleep, among other benefits.[7] Spending time to think about the good things in your life helps you change your internal dialogue. It can help shift the focus of your thoughts. The more you do it, the easier it becomes to think about things to be grateful for in whatever situation you might find yourself.

If you are more inclined to keep a regular journal instead of a gratitude journal, you will still reap many benefits, including managing anxiety, reducing stress and coping with depression.[8] Journaling can help you put your problems, fears and concerns in context and help you figure out what is rational and what might be overblown. It can help you track your mood to identify triggers. Personally, I found that it has helped me determine whether I became mad and angry about something for a good reason or whether I was already feeling that way and just blaming a particular event for my mood. Building up this awareness has been a useful tool in helping manage my depression. Now when I get mad or irritated at something, it is easier for me to let it go and not dwell on it if I know that my emotional reaction has more to do with myself than whatever just happened.

Meditation

Humans evolved a "fight or flight" response as an effective way to deal with real physical threats. However, our bodies are now reacting with this

---

[7]Jessen, L., The Benefits of a Gratitude Journal and How to Maintain One, HUFFPOST, Jul. 8, 2015. Available at: www.huffpost.com/entry/gratitude-journal_b_7745854 (Accessed on: Dec. 31, 2021)

[8]Journaling for Mental Health, University of Rochester Medical Center, Available at: www.urmc.rochester.edu/encyclopedia/content.aspx?ContentID=4552&ContentTypeID=1

response to everyday stresses such as money problems, bad drivers, job worries or relationship problems, which can be harmful to our bodies. This perpetual state of alarm can increase blood pressure, suppress the immune system and contribute to anxiety and depression.[9] Eliciting your body's relaxation response through activities like meditation and yoga can help you counteract the "fight or flight" response. Breath focus is a common feature of several techniques that evoke the relaxation response. Deep abdominal breathing can slow the heartbeat and lower or stabilize blood pressure.

Whatever technique you practice, from mindfulness meditation to yoga, tai chi, and Qi Gong, regular practice is essential. It helps enhance the sense of ritual and establish a habit. When you experience something scary or stressful, you can take a deep breath to reduce your stress level before you react.

Nature Therapy

In addition to the fact that sunlight can ease depression, especially seasonal affective disorder (SAD), being outside in nature has been found to have other benefits. One U.K. study found that a walk in the country helped reduce depression in 71% of participants.[10] The researchers found that as little as 5 minutes in a natural setting, whether walking in a park or gardening in the backyard, improves mood, self-esteem and motivation. The practice of shinrin-yoku or "forest-bathing" in Japan continues to grow in interest as it has been shown to promote health and happiness.[11] The term Nature Deficit Disorder reflects how our kids' focus on technology instead of nature has led to increased childhood obesity, attention deficiencies, early mental stressors and anxiety issues.

When my wife and I were operating a U-pick strawberry farm, the task of hand-weeding a couple of acres of strawberries often fell on me. When looking at row upon row of weedy strawberries, I sometimes wanted to quit.

---

[9]Relaxation techniques: Breath control helps quell errant stress response, Harvard Health Publishing, Jul. 6, 2020. Available at: www.health.harvard.edu/mind-and-mood/relaxation-techniques-breath-control-helps-quell-errant-stress-response

[10]Laguaite, M., Do You Need a Nature Prescription, WebMD, Apr. 13, 2021. Available at: www.webmd.com/balance/features/nature-therapy-ecotherapy#1 (Accessed on: Dec. 31, 2021)

[11]Walsh, J., McGroarty, B., Prescribing Nature, 2019 Wellness Trends from the Global Wellness Summit.

When I started to feel this way, I would sit back and take a few minutes to enjoy the mountains' beauty that surrounded me. Focusing on the fact that I was in one of the most beautiful places in the world would help make the task of pulling weeds more enjoyable.

Counseling

Sometimes all of the different self-help techniques just aren't enough to get us out of the hole we find ourselves in. Getting help from a counselor can help in many situations, including anxiety, depression, addictions, family and couple's issues, grief, etc. Benefits include:[12]

1. Increased skills in interpersonal communication.
2. Improved interpersonal relationships.
3. Decreased depressive symptoms.
4. Decreased anxiety symptoms.
5. Reduction in pharmaceutical interventions.
6. Improved quality of life.
7. Clarity of behavioral contribution to wellbeing.
8. Reduction of suicidal ideation.
9. Improvement in emotional self-regulation.
10. Reduction in substance misuse.

Once you think of action items and goals that would interest you, include them in your Personal Resilience Worksheet, which you can either download electronically at www.remicharron.ca/book or use the sheets provided in the Appendix. The following is an example of a plan for improving your mood.

Improving your mood	
Start this year	Follow through on diet and exercise activities
	Go for an hour walk in the woods every other weekend
	Start journaling once a week.
Start next year	Same activities and add: Spend 3 minutes a day reading
	jokes that make me laugh out loud.
	Finish each journal entry with three things I am grateful for.
Medium term, years 3–10	Same activities and add:
	Meditate for 15 minutes/day
Long-term, over 10 years	Same activities but increase meditation to 30 minutes/day

---

[12]Miller, K., 27 Scientifically Proven Benefits of Counseling, PostitivePsychology.com, Jun. 12, 2021. Available at: https://positivepsychology.com/benefits-of-counseling/ (Accessed on: Dec. 31, 2021)

6.2 Improving Your Psychological Resilience

"Anyone can be angry–that is easy. But to be angry with the right person, to the right degree, at the right time, for the right purpose, and in the right way–that is not easy."
– Aristotle

The exercises in the previous section for improving your mood will help you be more resilient. However, there are courses and books and activities specifically devoted to helping you become more psychologically resilient. This section will explore a few different exercises that you can do to help cultivate these skills.

Learning Your Emotions

Emotional intelligence is the capacity to be aware of, control and express one's emotions. This is a skill that I struggle with. I am often aware that I am having feelings but then struggle to describe these or figure out why I feel those emotions. Becoming more aware of our emotions can help us better manage how we deal with stressful situations. As an exercise, review a list of the different human emotions and if you aren't exactly sure what they mean, take the time to read their definitions. Paul Ekman, a psychologist and leading researcher on emotions, breaks down emotions into five categories: anger, fear, sadness, disgust and enjoyment (see following table).[13]

As a second step, reflect on what you are feeling right now. Write it down by naming the emotion. Thirdly, once you get better at putting into words your present feelings, you can take a day where you pause throughout the day to reflect on your emotions. After doing this for a few days, you can start to see patterns: what emotions dominate and what causes them? What are their triggers?

Being aware of your feelings can be difficult to master but it is useful. If I am already feeling irritated and someone does something that annoys me, I often get mad at them. However, I am not really mad at them for what they did, I simply assigned my feelings of irritation to what they did, even though

---

[13]Raypole, C., Big Feels and How to Talk About Them, healthline, Sep. 10, 2019, Available at: www.healthline.com/health/list-of-emotions (Accessed on: Dec. 31, 2021)

Anger	Fear 1. worried	Sadness
1. annoyed	2. doubtful	1. lonely
2. frustrated	3. nervous	2. heartbroken
3. peeved	4. anxious	3. gloomy
4. contrary	5. terrified	4. disappointed
5. bitter	6. panicked	5. hopeless
6. infuriated	7. horrified	6. grieved
7. irritated	8. desperate	7. unhappy
8. mad	9. confused	8. lost
9. cheated	10. stressed	9. troubled
10. vengeful		10. resigned
11. insulted		11. miserable
Disgust	Enjoyment	
1. dislike	1. happiness	
2. revulsion	2. love	
3. loathing	3. relief	
4. disapproving	4. contentment	
5. offended	5. amusement	
6. horrified	6. joy	
7. uncomfortable	7. pride	
8. nauseated	8. excitement	
9. disturbed	9. peace	
10. withdrawal	10. satisfaction	
11. aversion	11. compassion	

those feelings were already there beforehand. If I am already aware that I am feeling sad or irritated, I can better manage my reactions. Practicing this has helped reduce the number of unnecessary fights I was having with my wife. Instead of getting mad at her, I stop and take a few deep breaths and try to keep my mouth shut.

Practice Thought Awareness

We all have our internal dialogue. What we tell ourselves may be wrong or inaccurate, but it has a considerable influence on us. It's okay to have negative thoughts but don't let these derail your efforts. To be more resilient, practice positive thinking and "listen" to how you talk to yourself and see how your thoughts impact your emotions and actions. Do you find your mind drifting back to the same negative thought? Do you criticize yourself, thinking that you are dumb, ugly, lazy, etc.? Try to be aware of these thoughts. The first step is to identify the thoughts that may be negatively impacting you. The second step would be to redirect your mind to something else, like you might redirect a misbehaving toddler. Third, think about positive attributes about yourself and repeat them to yourself throughout the day. When you find yourself thinking negative thoughts about yourself, practice your positive mantra. The idea is to replace negative thoughts with positive, more realistic thoughts.

Lucy Hone, the author of "Resilient Grieving", suggests that we can develop more thought awareness into our everyday decisions by asking ourselves one simple question with each decision we make:[14] "Is this helping me or harming me?": That third glass of wine, does it help or harm? How about continuing to scroll through the news and social media or going for a walk? The question is a simple framework within which to take better care of ourselves. It adds an element beyond simple thought awareness to an awareness of the everyday decisions that we make.

Project Into the Past and Future

In psychology, mental time travel is a distinctively human skill. It involves rewinding to remember the past and fast-forwarding to envision the future.[15] If you find yourself struggling at the moment, if you focus on how awful your situation is, not only will you feel worse, you will also find that time goes by much slower. As an exercise, imagine strapping yourself into a time machine and traveling ahead to a day when your troubles have been resolved. For example, when in lock-down because of COVID, you could imagine the first vacation you would take once you could freely travel again. You might think that imagining this future might make your present situation more depressing, but psychologists have found that it actually makes us more cheerful.

We can either travel to the past to an event or time when we were happy or imagine a time in the future when things are good again. Dreaming like this reminds us that we're loved and helps us savor the joys of life. While traveling to the past, you can both relish in good memories or go back to difficult times. Remembering when things were worse helps us be more

[14]Holland, E., In a Crisis We Learn From Trauma Therapy, The New York Times, Jun. 6, 2020. Available at: www.nytimes.com/2020/06/15/health/resilience-trauma-emdr-treatment.html (Accessed on: Dec. 31, 2021)

[15]Grant, A. To Build Resilience in Isolation, Master the Art of Time Travel, The New York Times, May 15, 2020. Available at: www.nytimes.com/2020/05/15/smarter-living/to-build-resilience-in-isolation-master-the-art-of-time-travel.html

grateful for what we have in the present and it can help us help us remember how we have struggled through adversity before and have found our way to happiness again. You can also travel to a different past, imagining yourself going through a similar but worse situation (e.g. the plague versus COVID) and be grateful that things aren't that bad today.

Exploring Past Resilience

To have made it to where you are today, you already will have had past experiences with resilience. You made it through difficult situations. You can learn from these times in your life that were particularly challenging or demanding. Taking the time to think about how you handled these situations and made it through on the other side can help you be better prepared for the next time you are in challenging situations. Take the time to go through and reflect on these past situations:

1. What was the situation?
2. What challenges did you need to overcome?
3. What about it was particularly challenging for you?
4. What actions and reactions did you have that didn't help and maybe made things worse?
5. What helped you get through the situation?
6. Knowing what you know today, what would you have done differently to make things easier on you?

When we aren't struggling and take the time to reflect on going through difficult situations, it will help us better prepare ourselves for going through the next challenge that presents itself.

Cultivate Forgiveness

Some people dwell on the past. The more baggage we are carrying from the past, the harder it can be to focus on the future. Cultivating forgiveness can be beneficial to your mental and physical health.[16] The first step is to clearly acknowledge what happened, including how it feels and how it's affecting your life right now. Then, you make a commitment to forgive, which means

---

[16]Newman, K., Five Science-Backed Strategies to Build Resilience, Greater Good Magazine, Nov. 9, 2016, Available at: https://greatergood.berkeley.edu/article/item/five_science_b acked_strategies_to_build\resilience

letting go of resentment and ill will for your *own sake*. That's important. Forgiveness doesn't mean letting the offender off the hook or becoming friends with them. It's about letting go of the anger that you are carrying around.

Letting Go of Anger through Compassion is a five-minute forgiveness exercise that could help you. Spend a few minutes generating feelings of compassion toward your offender; she/he is only human and makes mistakes; he/she also has room for growth and healing. Be mindfully aware of your thoughts and feelings during this process and notice any areas of resistance. Cultivating compassion instead of ruminating on negative feelings or repressing them can help you build more empathy, positive emotions and feelings of control.

The above are just some of the exercises that you can do to start building your psychological resilience. You might not be interested in these particular exercises. If so, you can do your own research in this area. Once you think of action items and goals that you would be interested in doing, you can include them in your Personal Resilience Worksheet, which you can either download electronically at www.remicharron.ca/book or use the sheets provided in the Appendix. The following is an example of a plan for improving your psychological resilience.

Improving your psychological resilience	
Start this year	Find and read a book about emotional intelligence
Start next year	Think of a positive mantra and start repeating it when I wake up, before I go to bed and when I get stuck on negative thoughts.
Medium term, years 3–10	Take time to think about past grievances that I am still holding onto. Evaluate how I can forgive and move on.
Long-term, over 10 years	Take a course on psychological resilience and practice the exercises that are recommended.

Just like good physical health, there is minimal downside to planning for and implementing emotional resilience. Being resilient in the face of adversity is not all up to chance. You can do things today that will allow you to be more resilient in the future. Not everyone could practice the level of resilience that Nelson Mandela demonstrated. Still, everyone can improve their emotional resilience so that they don't get overwhelmed and give up at the first seemingly insurmountable challenge they face.

Now that you can plan for physical and emotional resilience, you can start to think of what skills and knowledge you can learn to build resilience.

7

Learning and Education – Kids and Homeschooling

"If children feel safe, they can take risks, ask questions, make mistakes, learn to trust, share their feelings, and grow."

– Alfie Kohn

Abstract

The second area of resilience that is covered in the book is Learning and Education. Again, this is broken down into two chapters, starting with a chapter on Kids and Homeschooling. This chapter provides some guidance to parents in terms of how they can help make their kids more resilient. For those parents that feel like homeschooling would be the best approach for their kids, the chapter starts by alleviating the two most common fears that parents have about homeschooling: that kids won't develop good social skills and that the parents don't have the skills to teach everything that is learned in school. It ends by describing a variety of resilience skills that you can help teach your kids outside of school.

Key Words: attachment parenting, homeschooling, social skills, resilience skills.

There is a growing number of people that are not having kids. For some, they are doing it as a way to reduce their impact on the climate and for others because they fear the world their kids will have to grow up in given the state of the climate. Although not having kids was something that we contemplated, my wife and I decided that having only two kids, less than the replacement rate (2.1) would be acceptable. Lower fertility rates can also lead to societal problems when you have fewer young people looking over a larger aging population. In terms of being scared for them to grow up in a world rocked by climate change, we felt that given the choice, they would likely rather live through it, than having not lived at all. That is why we made a conscious effort to raise them to become more resilient.

For those readers who have or are contemplating having kids, having them grow up to be resilient to deal with the challenges that life throws their way successfully is a universal goal. Depending on their age, working together to develop a Personal Resilience Workbook such that you can have both individual and family goals would be an excellent place to start. If you don't have kids and aren't planning on having any, feel free to skip to the next chapter.

Parents have a significant role to play in having their children grow up to be resilient, and it does not require them to have a lot of resilience training to teach this. How loved you felt as a child is a great predictor of how you manage all kinds of difficult situations later in life.[1] According to a review of the past 50 years of resilience research, the most important factor in developing resilience is the quality of our close personal relationships, especially with parents and primary caregivers. Early attachments to parents play a crucial, lifelong role in human adaptation.

This is not a parenting book and I will not detail how you can develop strong, healthy attachments with your children. Fortunately, "Attachment Parenting" is a parenting philosophy centered on creating a strong, healthy attachment. My wife and I have been following the attachment-based developmental approach developed by Dr. Gordon Neufeld. I encourage you to read one of his books or take some of the Neufeld Institute training (www.neufeldinstitute.org). We did not decide to follow attachment parenting for resilience. We just felt that the approach developed by Dr. Neufeld is centered on the latest research aimed at the wellbeing of the

---

[1]Zimmerman, E., What Makes Some People More Resilient than Others, The New York Times, Jun. 18, 2020, Available at: www.nytimes.com/2020/06/18/health/resilience-relations hips-trauma.html

child. Unfortunately, it is not a how-to book to make parenting easier as many parenting books you find available these days. It is about the wellbeing of the children and not necessarily the parents.

In this chapter, I will look at things we can teach our kids to make them more resilient. But before getting into the "what", I want to start with the "how". One thing you should not do is rely on the school to teach your children how to be resilient. A few school-based programs teach useful skills to children, such as gardening, but for the most part, we need to supplement what our kids are learning. School is not preparing our kids for the future. It is a very bureaucratic system that is slow to evolve. It has yet to catch up to the new reality of all kids having their own smartphones with them all the time, nor is it getting them ready for a world that will be radically different with many emerging technologies like artificial intelligence that are set to cause drastic changes, let alone a world with a destabilized climate system.

With COVID 19, many parents found themselves homeschooling without ever having signed up for it. Some parents hated every moment of it, whereas others enjoyed having the opportunity to be more involved in their children's learning. If you found yourself liking homeschooling or not entirely hating it, then you may want to consider homeschooling. Maybe just at the elementary level when they are growing up quickly so that you can focus on raising more resilient children. I am by no means saying that homeschooling is the only way. If you don't have the means to have a parent at home to homeschool or if you don't have the patience to do it while maintaining a strong loving attachment with your kids, then, by all means, send them to school. However, do think about how you will introduce some resilience elements into their education.

My wife and I have homeschooled our kids for most of their elementary school and have sent them to high school starting in grade 8. After a month of school, we pulled our daughter from kindergarten and our son from grade 2. When they started kindergarten, both of them were having some difficulties at school, which they were taking home with them. This was starting to impact their behavior and their attachment to us. Homeschooling definitely wasn't the easiest route, but we feel it was the right decision. When our son went back to school in grade 8, one grade ahead of where he was when we pulled him out, he wasn't behind academically, and a number of his teachers commented on how well-rounded he was compared to other kids his age. There are two myths that I want to dispel in case they are one of the factors keeping your kids at school: Myth 1 is that children need the social part of the school and Myth 2 is that children learn a lot at school.

7.1 Myth: Children Need School to Develop "Social" Skills

> *"Bullying builds character like nuclear waste creates superheroes."*
> – Zack W. Van

If you are worried that keeping your kids at home will deprive them of learning healthy social skills, take some time to think about how kids learn social skills at school. Generally, it happens before and after school, during recess and at lunchtime, when kids interact together without any adult supervision. They are learning social skills from each other before ever having learned how it's done. It is essentially the blind, leading the blind. When they are in class, they are with 20 plus kids of the same age, day after day, where peer pressure is enormous. Kids feel pressure to belong. They try to look and sound and be like everyone else before discovering who they are. So both structured and unstructured interactions often result in rivalry, ridicule, bullying, unhealthy competition, etc.

Depending on where you fall in the pecking order at school, it might be the best time of your life or it can be the worst. People that face bullying every day at school may only get away from the abuse once they are done school and go into the "real world", where normal school behavior may no longer be socially acceptable.

Dr. Richard Medlin, an expert in homeschooling from Stetson University, reviewed the literature on *Homeschooling and the Question of Socialization.*[2] His conclusions should help alleviate the fears on the socialization question:

> *"Compared to children attending conventional schools, however, research suggests that they [homeschooled children] have higher quality friendships and better relationships with their parents and other adults. They are happy, optimistic and satisfied with their*

---

[2]Medlin, R., Homeschooling and the Question of Socialization Revisited, *Peabody Journal of Education* **88** (3), 2013.

lives. Their moral reasoning is at least as advanced as that of other children, and they may be more likely to act unselfishly. As adolescents, they have a strong sense of social responsibility and exhibit less emotional turmoil and problem behaviors than their peers. Those who go on to college are socially involved and open to new experiences. Adults who were homeschooled as children are civically engaged and functioning competently in every way measured so far. An alarmist view of homeschooling, therefore, is not supported by empirical research. It is suggested that future studies focus not on outcomes of socialization but on the process itself."

Homeschooled children are not locked up at home with no social interaction with other kids their age. There are so many opportunities for them to interact with other kids, whether it be with organized activities with other homeschool kids, sports, art classes, music, they will get a chance to interact with other kids. However, a lot of these interactions are structured and have an adult teaching and leading them. Homeschoolers generally have more one-on-one interactions with adults. With more exposure to adults and structured interactions with other kids and less unsupervised, unstructured children-only group play, it is no wonder that homeschool kids can develop healthier social skills. This does not mean that they won't have friends with which they can have unstructured playtime, there is just less of it and generally, it will happen without some of the unhealthy group dynamics.

Having kids learn social skills from each other and having them spend so much of their time together leads to kids becoming peer-oriented, causing a host of issues at school and home. Peer-orientation is when children and youth look to their peers for values, identity and codes of behavior. This undermines parents and family cohesion, stunts healthy development, poisons the school atmosphere and leads to an aggressively hostile and sexualized youth culture. To learn more about this, I encourage you to read "Hold Onto Your Kids" by Gordon Neufeld and Gabor Maté.[3] It is a very important topic for resilience, given the evidence that healthy relationships with parents and caregivers are critical in being resilient in adulthood.

---

[3]Neufeld, G., Mate, G., Hold On to Your Kids: Why Parents Need to Matter More Than Peers, *Vintage Canada*, 2005

7.2 Myth: Children Learn a Lot at School

The second myth that I want to dispel is that kids learn a lot in school, and that it would be a challenge to cover all of the material they need to learn. Teachers spend a lot of their time in elementary school managing kids, often with a few in the class demanding most of their attention. After 8 years of homeschooling, I was surprised at how little time we needed to devote to the regular school curriculum to keep up with our peers. For the first few years, we pretty much practiced "unschooling" focusing on teaching them based on their interests. We then started workbooks to make sure that they covered the required curriculum. However, it did not take long to get through it every year.

We live on a farm and the active schooling often started later in September and ended sometime in May when things on the farm got busier. My wife would spend a couple of hours a day on the curriculum and manage to get through it just fine. Not to say that the kids didn't learn outside of this time; it just wasn't the regular curriculum. They spent a lot of time reading and doing things they enjoyed. For this to work, screen time needs to be severely limited. If not, kids won't generally develop many interests other than what revolves around their screen.

When homeschooling, it is easier to tailor what and how you teach your kids based on their individual learning styles and interests. This book that you are reading was part of a homeschooling challenge with my 12-year old daughter. She would often start writing books out of interest, without ever finishing them. For the 2020-21 school year, I challenged her to write and finish a book, agreeing that I would write one at the same time. This was an excellent way to give her the motivation to get it done. On the other hand, there is no way that I could convince my 14-year-old son to write a book. To get him to write, it needs to be shorter pieces that are focused on his interests like snowboarding or mountain biking.

Another problem with school is that they often try to teach some material too early. For example, some kid's brains are just not ready to learn to read in kindergarten. Teaching them is difficult and frustrating and these kids often fall behind if they are in school. However, if you had waited 1 year, they could have learned to read with a fraction of the time and effort. Finland, which has students with some of the best learning outcomes in the world, focus on play in kindergarten and only start teaching kids to read at age 7.

The same is true with some math concepts. My dad was a math teacher. He knew that some concepts such as negative numbers or fractions were much easier to teach if he simply waited later in the year for the kids' brains to be more developmentally ready. It could be physical activities too. I tried to teach my son to ride a bike without training wheels one autumn without success. He picked up his bike without training wheels in the spring and had it on the first try.

7.3 Teaching Resilience Skills

"The more that you read, the more things you will know. The more that you learn the more places you'll go."
– Dr. Seuss

You do not have to spend too much time teaching resiliency skills to your kids. As indicated, devoting time to developing healthy relationships with your kids is critical. That being said, there are some skills that you can teach your kids that will be useful, whether you do it as part of homeschooling or after school. The following are some resilience skills you can teach children.

Problem Solving: Whether your kids are problem-solving academically in math or science, while playing a board game such as Mastermind, Clue or computer adventure games, learning how to tackle problems effectively is a great skill to have. If life throws you a curveball, you can use it as a problem to apply your problem-solving skills, instead of getting overwhelmed with the challenge.

Learning from Mistakes and Failure: We all make mistakes and fail in life. Instead of getting angry and punishing your kids for their mistakes, teach them that mistakes and failures are inevitable and very useful learning tools. First, teach your kids not to be embarrassed about getting something wrong and then teach them how to reflect on these mistakes and failures to imagine how they would do things differently next time. Just practicing to fail in a safe environment is useful. My son would get really mad and frustrated when he lost at a board game. The more we played, the more he could practice losing without getting angry. Now he takes losing in stride.

Healthy Self-talk: Negative self-talk is unhealthy for anyone. It is much easier to get in the habit of developing an awareness of your self-talk and focusing it on positive thoughts when we're young. We sometimes get a glimpse of negative self-talk when kids blurt out things like "I'm so stupid" or "I'm a loser" or "I'm so fat", etc. If they are saying it out loud, it is likely part of their self-talk. Get them to recognize when they are thinking these unhealthy thoughts, and to practice some alternative things to say such as "I'll do better next time", "I like how I look" or "I have other great friends".

Facing Fears: Facing fears head-on is a great way to gain confidence. If you learn to not only confront your fears but to move beyond them, it can help you move forward when you are faced with a fearful situation. One way to help teach this is to get your kids to step outside of their comfort zone and face their fears one small step at a time. It might take some practice to help your kids face their fears with loving support instead of unhealthy pressure.

Practice Gratitude: Learning to be grateful for what you have, regardless of your situation, will not only make you happier, it will help make you more resilient. There are different ways to practice this. You can encourage your kids to keep a gratitude journal, where they write daily or weekly about what they are grateful for. You could also just take turns at the dinner table to say one thing that you are grateful about from your day.

Comfortable with Emotions: Teaching your kids from a young age what the different emotions are named and allowing them the space to feel their emotions is important. Many parents start to feel uncomfortable with the more negative emotions such as anger or sadness. As a result, they may try to stop kids from expressing these feelings. Don't stop them from expressing these emotions, just steer them in healthier ways of doing it if it is causing harm to themselves or others. You can't stop them from feeling these emotions and forcing them to internalize these feelings is not healthy.

Broader Set of Skill Development: In general, the more skills your kids know, the more resilient they will become. You never know when a skill might become handy in getting you out of a bad situation. There are many essential life skills that school does not teach that are important to learn before adulthood such as cooking, cleaning, first aid and personal finance. On top of these life skills, you might also want to teach your kids other skills that can be handy for them to have in life, such as how to grow their own food, basic wilderness survival skills, sewing, computer programming, small engine repair, etc. Again, you might not want to cover all of these and it might be useful to focus on things that are more aligned with your kids' interests, but teaching practical skills that they can use in life should be part of your kids' education. Having our kids grow up on a farm so that they would gain a broad set of life skills was one of the reasons we moved in with my in-laws on their farm. Even without forcing them to learn, they have picked up a lot of knowledge that may become useful to them as adults.

Once you think of action items and goals that you would be interested in doing, you can include them in your Personal Resilience Worksheet, which you can either download electronically at www.remicharron.ca/book or use the sheets provided in the Appendix. The following is an example of a plan for your kids' learning and education.

Your kids' learning and education	
Start this year	Kids' cook one meal per week Start saying one thing we're grateful for after dinner
Start next year	Start an annual family camping trip Plant and take care of a vegetable garden as a family
Medium term, years 3–10	Organize a week-long canoe-camping trip for the family Take a skill-building course together at the local community center
Long-term, over 10 years	Continue annual camping trips as a family but change them to hike-in or canoe-in camping trips

As much as homeschooling or schooling at home will be a learning experience for the parents and the kids, the parents should also be on a lifelong learning journey building their skills and knowledge, which brings us to the next chapter.

8

Learning and Education – Personal Skills and Knowledge

"Live as if you were to die tomorrow. Learn as if you were to live forever."
– Mahatma Gandhi

Abstract

The second chapter on Learning and Education covers skills and learning activities that the reader can do to become more resilient. This is broken down into three categories, each with its own sub-heading: Career skills, Skills to increase your resilience and General knowledge building. In the end, the importance is to keep learning, as any new knowledge you gain can help you on your journey to increased resilience.

Key Words: resilience skills, career skills, knowledge building, lifelong learning, professional certification.

Lifelong learning is not just a slogan, it is essential to practice this more than ever. Technological change is happening at an exponential rate. The rate of change keeps increasing such that in a few decades, the world will be unrecognizably different than it is now. According to Ray Kurzweil, a director of engineering at Google, "We won't experience 100 years of progress in the 21st century – it will be more like 20,000 years of progress"

71

compared to the past.[1] Similarly, climate change and other forms of environmental degradation are also happening at an exponential rate, where every year will typically see a larger change than the last. If you are not keeping up with this rapid change, life will speed past you.

To keep up with all of this change, you need to develop a growth mindset. Some people have a more fixed mindset, believing that human qualities are innate, that you are either born with them or not. In comparison, a growth mindset maintains that human qualities are adaptable, that things can be developed or changed over time. Carol Dweck has developed the concept of the growth mindset and has shown that merely believing that you can keep improving enables you to believe that you are in control of your skills and that you can learn and develop new skills and improve them throughout your life.[2] The idea that you can learn and grow as an individual is pivotal in developing your Personal Resilience Plan. It is essentially a growth plan for you to learn and become more resilient over time.

In addition to thinking about your kids' learning and education, which was covered in the last chapter, this chapter focuses on your personal learning and education goals. It is divided into three parts. The first covers the need to keep building new skills for your career, the second focuses on learning skills that will make you more resilient and the third focuses simply on the benefits of learning anything.

8.1 Career Skills

"Move out of your comfort zone, develop those necessary skills and go all out for that much needed advancement!"

– Abhishek Ratna

Gone are the days when you can go to school, learn your trade, and get one job to apply your skills for your whole career. With the average person going through 5-7 career changes and 12-15 positions, learning new skills and building more knowledge will make you more resilient on the job front.

---

[1]Butler, D., Tomorrow's World, *Nature* **530**, Feb. 25, 2016

[2]Decades of Scientific Research that Started a Growth Mindset Revolution, mindset works, Available at: www.mindsetworks.com/science/ (Accessed on: Dec. 31, 2021)

Although you can learn new skills after losing your job, it is always a good idea to keep learning so that you are ready to look for a different career/job if needed. It is also not just a matter of learning skills to get a new job. You need to keep learning just to keep the job that you have in order to keep up with the pace of change. That is one reason why more and more professional associations are implementing mandatory professional development hours for their members to keep their certifications/designations.

Some skills are transferrable from one industry to the next, such as project management, human resources, communications, etc. You can enhance these skills by tailoring any potential professional development opportunities at work and taking courses to get you a certification. Many professional associations offer courses and certifications that could help make the skills you learn at one job more transferable to another. Examples include certified Project Management Professional[3], certified Human Resource Professional[4], Communication Management Professional[5], etc. Since I was doing a lot of work in building energy efficiency, I took a course and got certified as a Passive House Designer. I am now in the process of taking courses to become a certified Infrastructure Resilience Professional. Neither of these certificates is required for my current jobs, but they do open doors to new consulting and job opportunities.

It is always a good idea to take advantage of all the learning opportunities available through work, whether it is courses that are offered at the office or that work offers to pay for you to keep learning. This might be especially useful if you are doing a very specialized skill in an industry that is struggling with the rapid pace of change (e.g. photograph developer, switchboard operator) or a profession that artificial intelligence will disrupt in the coming years (e.g. banking, financial services, insurance, legal).

If you are new to the fossil fuel industry, there is a high likelihood that increased global efforts to mitigate climate change will impact your job at some point in your career. As with most jobs in the resource sector, it pays well during boom times, but busts can lead to large-scale layoffs that can

---

[3]PMI Certifications, Project Management Institute, Available at: www.pmi.org/certificatio ns

[4]Certified Human Resource Professional, Human Resources Professionals Association, Available at: www.hrpa.ca/designations/chrp-certified-human-resource-professional/

[5]Certification Programs, Global Communication Certification Council. Available at: https://gcccouncil.org/About-GCCC-Certifications

make finding another job difficult. It can be even worse if you are in a small industrial town, and the primary industry shuts its doors. If you have doubts about how long you will still have a job and that you think it would be difficult to find something comparable, you should have a plan on how you would transition from your current job to a new one, if required.

If you think you would have a challenge finding another job if you lost your current one during a downturn, you can take college or university courses or programs to learn something new. A new degree or certificate can help you pivot to a different sector. You might take trades-related courses such as plumbing to learn how to maintain your house, which could help you get work as a handyperson in the future. Many people earn their MBA's to find a job in management in a different job sector. It can just be something you do out of personal interest that you can have as an option in your back pocket or something that opens doors for you as a preemptive move. Granted, earning a certificate or degree can be a significant financial and time commitment that not everyone has. It could be that you just do research now into potential certificates or degree programs that you would be interested in doing if you are laid off. If I needed to learn a new skill to change jobs, I would either do a Certificate in Data Analytics or a Certificate in Artificial Intelligence. Both are of interest to me if I had the time or the need.

Once you think of action items and goals that you would be interested in doing, you can include them in your Personal Resilience Worksheet, which you can either download electronically at www.remicharron.ca/book or use the sheets provided in the Appendix. The following is an example of a plan for gaining career skills.

Career skills	
Start this year	Research professional certificates available that I could qualify for given my current job. Find out what policies work has regarding support for professional development.
Start next year	Find and take a course that will improve my worth at my current job and create opportunities for a future job switch.
Medium term, years 3–10	Continue taking at least one course per year towards my career.
Long-term, over 10 years	Pursue jobs in management in my field.

8.2 Skills to Increase Your Resilience

> *"Resilience isn't a single skill. It's a variety of skills and coping mechanisms."*
> – Jean Chatzky

Whether it is because of a major storm that caused regional havoc, a global shutdown due to a pandemic, a major economic depression following a pandemic or more personal struggles with job loss, there will be times when local goods and services might be disrupted, unavailable or simply just too expensive for you. If you are someone that relies exclusively on other people to grow your food, fix everything, cook your meals, etc., you can find yourself in a tough situation. As COVID 19 demonstrated, if there is a disruption in supplies or even the rumor of disruption, it doesn't take long before panic buying and hoarding ensues. All of the toilet paper in the region has disappeared before you know it, and you're stuck using reusable cloths.

My wife loves kombucha (bubbly fermented tea). She would easily buy 4 to 6 liters a week from the health food store, which is relatively expensive. She had just finished learning how to make her own kombucha as a way to help save some money when COVID 19 hit. For months our local kombucha supplier stopped delivering to our health food store. My wife was thrilled that she could make her own and not rely on someone else.

Becoming more resilient can mean learning skills that can make you less dependent on others to fulfill your daily needs. You can also learn some skills that would be useful in helping you survive (e.g. first aid, foraging, etc.). In times of need, you can not only use some of these skills to help you save money but they can also potentially be used as a way to make a bit of extra cash. Your skills can also be used as a bartering tool with your neighbors, where you help each other with what you are good at. Or as the case with the kombucha with my wife, it can become your only means of getting something. The following are some examples of skills that you can learn to increase your resilience:

1. Sewing and mending your clothes.
2. Cooking healthy, tasty meals.
3. First aid skills.
4. Wilderness survival skills.

5. Small appliance, engine or auto repair.
6. Basics of carpentry, plumbing, welding.
7. Basic IT skills.
8. Hunting and fishing.
9. Making beer, wine, cider, kombucha or alcohol at home.
10. Foraging for wild food.
11. Growing and processing your own food (veggies, meat, preserves, etc.).

Even though it is already a long list, I am sure that you could think of more skills to add to this list. Focus on areas that you are more interested in learning. You could also assess your situation, determine what you couldn't do without (e.g. a cold beer) and learn the skills you would need to look after yourself. As part of this analysis, you can also assess your close family and friends' skills that you will likely be able to rely on in times of need. For example, my brother-in-law, who lives on the farm with us, is a welder and good at fixing things. Instead of learning these same skills, I can focus on others we currently don't have on the farm.

Once you think of action items and goals that you would be interested in doing, you can include them in your Personal Resilience Worksheet, which you can either download electronically at www.remicharron.ca/book or use the sheets provided in the Appendix. The following is an example of a plan for learning skills to increase resilience.

Skills to increase resilience	
Start this year	Make some dandelion wine Preserve (can) some vegetables for winter (carrots, beans, asparagus)
Start next year	Hunt for a deer and butcher it myself Learn more local spots to go fishing Preserve (can) tomato sauce
Medium term, years 3–10	Learn how to smoke fish and other meats Learn how to ferment foods for preservation Learn how to graft grapes to start growing varieties for wine
Long-term, over 10 years	Learn how to make wine

8.3 General Knowledge Building

> *"An investment in knowledge pays the best interest."*
> – Benjamin Franklin

It appears that learning anything can help us on our quest to become more resilient. According to a 2020 article in the New York Times, *"the quest to understand something new is a key factor to building the resilience necessary to weather setbacks and navigate life's volatility"*.[6] Essentially, if you are continuously learning new things, you build confidence in your ability to figure things out once a crisis hits. To effectively deal with change, it helps to be engaged in changing yourself. The article quotes Dorie Clark of Duke University *"it is about having the kind of cross-training necessary to be flexible in an uncertain world where we don't know what is around the corner"*.

When it comes to learning for fun, you should follow your passions. If you like to cook, take cooking classes, if you want to play the guitar, take guitar lessons, etc. If you follow your passion and learn more about what you love, it could even one day turn into a career. You often hear stories or anecdotes of people that turn their hobby or interest into a career. This is not for everyone, because what you found fun and relaxing as a hobby can turn stressful as a job. The novelty of doing what you love can soon wear off when you're forced to do it. It does not have to be all or nothing. Some people turn their hobbies into a side-gig where they can earn extra money on their own time, which could be something you could fall back on if needed.

You don't need to worry about what to learn and what to do with that knowledge. Just take the time to learn something that is of interest to you. I like to read novels, but every couple of books I read, I try to read a nonfiction book. Just reading a nonfiction book means that I am learning new things. If not books, you can take courses in your local community or look online at the many different places that offer free or reasonably priced online classes, such as Coursera, EdX, The Great Courses, LinkedIn Learning, MasterClass, Skillshare, TED Talks and Udemy. There are even a

---

[6]Hannon, K., To Build Emotional Strength, Expand Your Brain, The New York Times, Sep. 2, 2020, Available at: www.nytimes.com/2020/09/02/health/resilience-learning-building-ski lls.amp.html

lot of YouTube videos online where people are teaching a skill they know. What things have you been meaning to learn but haven't made the time?

Once again, when you think of action items and goals that you would be interested in doing, you can include them in your Personal Resilience Worksheet, which you can either download electronically at www.remicharron.ca/book or use the sheets provided in the Appendix. The following is an example of a plan for general knowledge building.

General knowledge building	
Start this year	Read one nonfiction book per month
	Take a photography class at the community center
Start next year	Continue reading non-fiction books
	Learn how to use photo-editing software to make photos more impressive
Medium term, years 3–10	Take an oil painting class
	Take the local weekend retreat where participants learn to paint
Long-term, over 10 years	Take mixed media art class to learn how to combine pictures and painting

9

Community Building

"As far as the business of solitary confinement goes, the most important thing for survival is communication with someone, even if it's only a wave or a wink, a tap on the wall or to have a guy put his thumb up. It makes all the difference."

– John McCain

Abstract

The third area of resilience that is covered in the book is Community Building. Essentially, we are more resilient when working together than when everyone fends for themselves. The chapter presents three different areas where you can work on community building: Neighbors, Local community and Friends and family.

Key Words: neighbors, social networks, local community, friends and family.

Humans are social beings. Having close friends and family is an essential part of what makes us happy. It is also an important part of resilience. If you have someone to fall back on in times of need, you are much more likely to get through whatever you are struggling with much quicker. In its resilience report, the American Psychological Association wrote: "Many studies show that the primary factor in resilience is having caring and supportive relationships within and outside the family. Relationships that create love

79

and trust provide role models and offer encouragement and reassurance, help bolster a person's resilience."[1]

One of the challenges of the COVID 19 lockdowns was this lack of social connections. This was particularly challenging for the elderly in care homes since social relationships are critical for older adults to cope with declining cognitive abilities and other health challenges. However, it is one time when social networking tools helped us stay connected even when we were self-isolated at home. Even without connecting, just knowing that there are people somewhere on this globe that are cheering for you and that you can reach out to is a driver of resilience.[2]

However, having a strong community is more than just having good relationships with our friends and family. Having close relationships with our neighbors and with the greater community will also help in building your resilience. In this Community Building chapter, we first look at the importance of building relationships with your neighbors, then the greater community and then finally at the importance of friends and family.

9.1 Neighbors

"Be at war with your vices, at peace with your neighbors and let every New year find you a better man."
– Benjamin Franklin

In the past, people were a lot more connected with their neighbors. There was less travel, no internet, your neighbors were a big part of your safety net, whether for a missing cup of sugar or to help you through hard times. People tended to stay in the same jobs and live in the same neighborhood for their whole career, making it easier to have closer-knit communities. Society has become more individualized in terms of how we choose to spend our leisure time. Everyone seems glued to their smartphones connected to social networks that are no longer defined by geography. Kids used to play outside

---

[1]The Road to Resilience, American Psychological Association, Washington, DC, 2014.

[2]Holland, E., In a Crisis We Learn From Trauma Therapy, The New York Times, Jun. 6, 2020. Available at: www.nytimes.com/2020/06/15/health/resilience-trauma-emdr-treatment .html (Accessed on: Dec. 31, 2021)

with the other kids in the neighborhood. Now, they are often over-scheduled with activities that aren't part of the neighborhood and when they are home, they are inside plugged into their screens.

You don't need to look far to find research that shows the importance of neighbors. Strengthening neighborhood-level relationships increases community resilience, specifically when it comes to emergency preparedness, disaster response and recovery. Past disastrous events, such as the Indian Ocean tsunami in Southeast Asia and Hurricane Sandy in New York, have highlighted that neighbors are a significant source of help during recovery.[3] Knowing your neighbors has many benefits beyond helping boost your resilience in times of need and can include:[4]

1. Feeling of belonging in your community, which is an important factor in improving your wellbeing.
2. Facilitate mutual aid and support between people.
3. Reduces crime and encourages people to call out bad behavior.
4. Increases your social network, which leads to more opportunities in life.

There has also been a trend of people moving away from traditional neighborhoods into cities and high-rise buildings. Unfortunately, despite increasing the number of neighbors, people living in these buildings often actively try to avoid each other. Research has shown that apartment dwellers experience higher social isolation levels from their neighbors than those living in townhomes or single detached homes and that renters have weaker connections with their neighbors than condo owners.[5] Luckily there are some initiatives underway to help neighbors get to know each other. The City of Vancouver launched the Hey Neighbor! Pilot project with the aim to:[6]

1. Increase a sense of community amongst residents within their buildings.

---

[3]Chia, E., Building Neighbourhood Social Resilience, University of British Columbia, Vancouver, 2014.

[4]Hothi, M., Cordes, C., Understanding neighbourliness and belonging, The Young Foundation, 2007.

[5]Chia, E., Building Neighbourhood Social Resilience, University of British Columbia, Vancouver, 2014.

[6]Hey Neighbor!, City of Vancouver, Available at: https://vancouver.ca/people-programs/hey-neighbour.aspx

2. Decrease the frequency and intensity of loneliness among residents.
3. Support participating buildings to feel like home and not just a temporary place of residence.
4. Increase sense of responsibility and care over common property amongst residents.

To do this, two or three residents of the building became resident animators (RAs) and organized events for everyone in the building. These events included board-game nights, potlucks, walking clubs, emergency preparedness workshops, skill shares, etc. The pilot was a success and the work to encourage more is ongoing through the Hey Neighbor Collective.[7]

Whether you live in a single-family house in the country or rent an apartment in the city, you should make an effort to get to know your neighbors. It can not only build a feeling of shared responsibility where you will help each other in times of need, but it will likely also bring benefits to your daily life. Once you think of action items and goals that you would be interested in doing, you can include them in your Personal Resilience Worksheet, which you can either download electronically at www.remicharron.ca/book or use the sheets provided in the Appendix. The following is an example of a plan to build your local community through your neighbors.

Community building – neighbors	
Start this year	Bring a jar of preserves to our new neighbors to introduce ourselves
Start next year	Organize a "work bee" weekend in the spring with neighbors where we help prepare the garden space for each of the participants
Medium term, years 3–10	Organize a running group with the neighbors Participate in the jam exchange where everyone trades their homemade jams to have greater variety
Long-term, over 10 years	Offer to hire the neighborhood kids to work on the farm as they get to working age.

---

[7] Hey Neighbor Collective, Simon Fraser University, Available at: www.sfu.ca/dialogue/programs/urban-sustainability/hey-neighbour-collective.html

9.2 Local Community

"One of the marvelous things about community is that it enables us to welcome and help people in a way we couldn't as individuals. When we pool our strength and share the work and responsibility, we can welcome many people, even those in deep distress, and perhaps help them find self-confidence and inner healing."

– Jean Vanier

If you expand your efforts beyond just getting to know your neighbors and strengthen your connections with the local community, that will help build your overall resilience. Strong and diverse communities will fare better through tough times. There is a lot of research done on ways that communities can improve their resilience.[8] However, this book's focus is on the steps you can take to improve your personal resilience. Therefore, the discussion will focus more on activities that you can personally do to build a local community.

Local businesses not only provide local jobs, they also provide you with buying choices. Having a few large corporations supplying us with the majority of our goods and services is not resilient. What happens if there is a prolonged strike at Amazon or Walmart or the main Port bringing goods from our international suppliers is shut down? There are some advantages of our hyper-globalized society, but it has come at the expense of making our local communities much less resilient.

Amazon fared exceptionally well during the COVID pandemic. Consumer spending on Amazon between May and July 2020 was up 60% from the same time in the previous year. It is predicted that in 5 years, e-commerce will make up one-quarter of total retail sales, up from roughly 15% in 2019, likely making Jeff Bezos, owner of Amazon, the first trillionaire in history.[9] This shift in consumer spending will lead to the closing of 100,000 brick and mortar retail outlets in the US alone. The convenience of online shopping is

---

[8]Patel, S., Rogers, M., Amlot, R., Rubin, G.J., What Do We Mean by 'Community Resilience'? A Systematic Literature Review of How It Is Defined in the Literature, *PLOS Current Disasters*, Feb. 1, 2017. Available at: http://currents.plos.org/disasters/index.html%3Fp=28783.html

[9]Semuels, A., Many Companies Won't Survive the Pandemic. Amazon Will Emerge Stronger Than Ever, *Time*, Jul. 28, 2020. Available at: https://time.com/5870826/amazon-coronavirus-jeff-bezos-congress/

incredible. Anything you can think of is a click away. But this convenience comes at a cost to the majority of the small business owners in your community. How many businesses in your community have disappeared for good as a result of COVID or with the proliferation of e-commerce?

One of the things you can do to help improve your personal and community resilience is supporting your local community businesses and volunteer organizations. Some examples include:

Local Farms: Support your local farmers by shopping at your local farmer's market, buying a farm share through a community shared agriculture (CSA) program, buying your meat direct from a farmer, etc. COVID 19 restrictions closed down many meat processing facilities, which led to some shortages at local grocery stores. Our farm sells pork, lamb and beef to regular local customers. The demand during COVID shot up. However, we can't produce animals out of nothing. We were able to sell to a few new customers, but we gave priority to our existing customers. When in need, it can pay off to have a long-standing relationship buying meat, eggs and produce from your local farmers.

Food Bank: Volunteering at any local organization is an excellent idea to increase your resilience, as altruism is beneficial to your overall wellbeing. Volunteering at a local food bank helps an organization that can be there in your time of need. If you don't have the time, you can always donate food directly or through most local grocery stores. COVID 19 was particularly bad for food banks, where demand for food assistance rose at an extraordinary rate while at the same time having less donated food and volunteer workers.[10] During the great recession of 2008, food banks across Canada saw a 28% increase in demand. Since you never know when you or your loved ones might be in need, it is a good idea to support institutions that provide a safety net for your community when you have the means to do so.

Participate in Seed and Preserve Exchanges: In my local community, some groups get together and exchange preserves that they have made. Everybody that goes to the meeting brings a set amount of jars of preserves (e.g. jams, salsas, pickled veggies, etc.). You then get to take home one or

---

[10]Kulish, N., 'Never Seen Anything Like It' Cars Line Up for Miles at Food Banks, *The New York Times*, Apr. 8, 2020, Available at: www.nytimes.com/2020/04/08/business/econom y/coronavirus-food-banks.html

two of all of the different jars that people brought. This brings people together, encourages them to preserve food and increases the variety of recipes you get to try. Similarly, seed exchanges work when people bring seeds of fruits and vegetables that they like to grow. This can help protect some heritage seeds and enables you to learn what varieties grow well in your local climate. Our local library has a bin where people can bring and take seeds to exchange. There are many seed exchange groups out there, such as Seed Savers Exchange in the US, which is dedicated to preserving and sharing heirloom seeds.[11] Join an existing exchange or start your own.

Local Restaurants and Retail: As discussed, there are many advantages to having strong local businesses in your community. Local businesses often support local charities. The business income is more likely to stay in the community, and the owners have a vested interest in serving the community. Multinationals can be quick to set up shop but can also disappear overnight as the conglomerates decide to restructure their businesses. Even though I can save a few bucks by shopping online or going to a nearby city, I like to buy local to support our local businesses. The small independently owned restaurants and retail stores provide municipalities with vibrancy and unique character. Small business owners are also more likely to build personal relationships with their customers, knowing many by name, creating a sense of community.

Once you think of action items and goals that you would be interested in doing, you can include them in your Personal Resilience Worksheet, which you can either download electronically at www.remicharron.ca/book or use the sheets provided in the Appendix. The following is an example of a plan to help build your local community.

Build your local community	
Start this year	Buy new shoes and clothes at the local general store
	Grow extra food and donate it to the local food bank
Start next year	Get a weekly food box through our neighbor's CSA program
	Continue growing food for the local food bank
	Buy new bikes for the kids at the local bike store
Medium term, years 3–10	Shop local for all Christmas gifts
	Continue getting food box through our neighbor's CSA program
Long-term, over 10 years	Continue shopping locally and providing food to the food bank
	Donate to other local charities

---

[11]Find and Share Seeds, *Seed Savers Exchanged*, Available at: https://exchange.seedsavers.org/

9.3 Friends and Family

"Nourishing relationships is the single most universally agreed-upon feature
of the good life."
– Daniel Goleman

Having close friends and family members that we can hang out with and talk to about our struggles is essential to our overall happiness and resilience. Research has found that one close friend can enable teenagers to bounce back quicker from stressful life events and to learn and grow from them.[12] The benefits of friendship will extend into adulthood. Unfortunately, as we get busier with work and family obligations, people often tend to put friendships on the back burner. Men are more at risk than women, with many having no close friends as they approach middle age.[13]

In addition to increasing our happiness and our ability to handle stressful situations, close family and friends can also add resilience by sharing their skills and experience during times of need. If you have a more extensive network of people that you can rely on, not everyone needs to be good at everything. One person might be a great cook, another person can grow food, someone can be good at fixing things, etc. If you can all rely on each other's strengths, it will make things easier.

When it comes to friendship, quality counts more than quantity. How many of your "friends" on Facebook or other social media would you open up to and share how you are feeling and what is stressing you out? While there can be some advantages of having a more extensive and diverse network of friends and acquaintances, it is more important to nurture a few truly close friendships that will be there for you through thick and thin. Besides, how many close relationships can you handle in your life? Social scientists and

---

[12]Evans-Whipp, T., Gasser, C., Teens with at least one close friend can better cope with stress than those without, *The Conservation*, Nov. 24, 2019, Available at: https://theconversation.com/teens-with-at-least-one-close-friend-can-better-cope-with-stress-than-those-without-126769

[13]Payne, E., The Importance of Resilience, *Payne Resilience Training & Consulting*, Available at: https://payneresilience.com/

psychologists who study relationships often suggest that most of us can only really manage 7 to 10 people in our lives at any one time.[14]

In your larger social network of friends and family, how many would you consider close friends? How many would you call about your problems or notice when they are struggling? Since many adults spend most of their time at work, it might be a good place to find a friend. However, we often mix up colleagues as friends. Friendship needs to be deeper than the superficial elements that we generally stick to with colleagues or other close acquaintances. It can be challenging to transition those relationships into friendships. I have tried and failed on several occasions. After you invite them over for dinner or other activities, you can soon realize that those efforts are one-sided. They are more interested in maintaining a "friendly" relationship that does not involve actually being friends.

Beth Payne, who provides resilience training and consulting, provides four steps to develop close friendships to help build your resilience:[15]

1. **Create New Connections:** If you don't have friends in your area, join a club, a meet up, a course or other activity. Look into volunteer opportunities with a local charitable organization. Getting out and meeting new people will help you expand your network of potential friends.
2. **Strengthen Existing Connections:** Many of us have old friendships that we either take for granted or have neglected over time. You might have even let the relationship with your spouse dwindle. If you are married or have friends but still feel lonely, this is an indication that your relationships might need strengthening. Reach out to these friends to connect over coffee or a meal and catch up. Consider carving an extra hour or two out of your week to regularly schedule lunch or dinner with a close friend.
3. **Spend Time in Person:** Research has shown that casual friendships emerge after only 30 hours spent together, whereas close friendships do not develop until after 300 hours spent together. While the internet can

---

[14] Albrecht, S., The Sad Reality of Friends Versus Acquaintances, *Psychology Today*, May 11, 2019. Available at: www.psychologytoday.com/ca/blog/the-act-violence/201905/the-sad-reality-friends-versus-acquaintances

[15] Payne, E., Four steps to developing close friendships, *Payne Resilience Training & Consulting*, Feb. 10, 2020. Available at https://payneresilience.com/blog/

be a powerful tool to stay connected, it is important to have plenty of face-to-face time to sustain and nurture relationships. Catching up, joking and having meaningful conversations are all important.

4. **When Stressed, Reach Out:** An essential part of building resilience is accepting help from the people who are important to you. If you are struggling with stress or a tough situation, lean on a friend for support. Asking for help is a great way to make yourself feel better and deepen your bond with that friend.

In addition to making new friends, it might be useful to actively think about what friends and family you can currently count on in different situations. In 2007, Great-West Life launched the Centre for Mental Health in the Workplace, now called the Workplace Strategies for Mental Health. As part of this initiative, they released an Employee Guide – Plan for Resilience.[16] In it, they stress that resilience involves acknowledging our interdependence with each other. They include an exercise where you write down who you can depend on for many different situations (e.g. help with housework, provide emotional support, make you laugh, motivate you, have fun, set you straight, etc.). This can help you know who to reach out do for different circumstances.

Once you think of action items and goals that you would be interested in doing, you can include them in your Personal Resilience Worksheet, which you can either download electronically at www.remicharron.ca/book or use the sheets provided in the Appendix. The following is an example of a plan for building stronger relationships with your friends and family.

Community building – friends and family	
Start this year	Start monthly date nights with my wife
	Call my siblings and parents every weekend
	Go on a weekend fishing trip with my friends
Start next year	As above and:
	Join the local running club to meet new friends
	Go on a monthly lunch with colleagues from work
Medium term, years 3–10	Go on a vacation to my hometown to spend time with a few old high school friends.
	Go out to supper once a month with a friend
Long-term, over 10 years	As above, and keep a close relationship with my kids as they move out of the house.

---

[16]Workplace edition -Plan for resilience, Canada Life, Available at: https://wsmh-cms.med iresource.com/wsmh/assets/h6ov7ka37jswo040 (Accessed on: Dec. 31, 2021)

10

Financial Preparation

"Looting speaks to a lack of economic opportunities – frankly, we all would loot, too, if our families' continued survival depended on it."
– Sarah Parcak

Abstract

The fourth category that is covered in the book is Financial Preparation, presenting how we can make ourselves more financially resilient. It has useful information for people that have little to substantial amounts of available money. It covers four different categories that the reader can focus on to improve their financial resilience: Savings, Debt repayment, Income generation and Protecting your assets.

Key Words: emergency savings, debt, budget, retirement, income, insurance, will.

If you have a lot of money, it can be much easier to plan to be financially resilient in many different situations. However, most of us don't have an overabundance of money. It becomes a matter of prioritizing what we do with the money that we have at our disposal. The less money you have, the smarter you have to be with that money. Unfortunately, the opposite is often the case. People with less money also tend to have lower financial literacy.[1]

---

[1]Lusardi, A., Hasler, A., Yakoboski, P., Building up financial literacy and financial resilience, *Mind & Society* **20**, 181-187, 2021

Being financially resilient includes understanding and managing risk. It turns out that when it comes to financial literacy, comprehending risk and insurance are areas where knowledge is lowest.

This chapter breaks down how you can prepare to be more financially resilient into four areas: Savings, Debt Repayment, Income Generation and Protecting Your Assets. If you don't understand much about finances, it would be a good idea to improve your financial literacy by taking a course or reading a book. Once you have read through this chapter, it might also be a good idea to consult with a financial planner to help execute your plan.

10.1 Savings

"Do not save what is left after spending, but spend what is left after saving."
– Warren Buffett

Having savings to fall back on during an emergency is an integral part of being financially resilient. As with over three-quarters of Americans, if you are up to your eyeballs in debt and living paycheck to paycheck, it does not take much of a setback before you are in trouble[2]. This section looks at savings by establishing a budget, saving up for emergencies and other future anticipated expenses and finally saving for retirement.

Review Your Budget: Everyone should have a budget to ensure that expenses remain in line with income and that your money is going where you want. Would you rather spend $5 per day on a fancy coffee on your coffee break or spend $1,200 per year on a vacation? The less money you make, the more careful you have to be that you aren't spending your money on frivolous things instead of things you actually want or need. The first step in creating a budget is to understand where all of your money goes.

Review your regular fixed expenses like rent, food, transportation, insurance, etc., and take a few months to track all of your spending. There are quite a few apps out there that can help you track your spending and help you

---

[2]Friedman, Z., 78% Of Workers Live Paycheck To Paycheck, *Forbes*, Jan. 11, 2019, Available at: www.forbes.com/sites/zackfriedman/2019/01/11/live-paycheck-to-payche ck-government-shutdown/?sh=5a8171304f10

develop an overall budget. Once you have a clear understanding of your current income and expenses, you can build your budget. As part of budgeting, you should look at all of the items in this chapter to see if you want to boost your savings, investments, debt repayment or insurance to improve your financial resilience. If you don't have enough income to meet all of your goals, you will need to examine your expenses to see where you can find savings. If you can't find a place to cut, consider how you could earn more income. Your final budget will need to prioritize where all of your money goes based on your needs, desires and goals.

Build Your Emergency Fund: A study found that 56.3% of Americans have less than $1,000 in their checking and savings combined.[3] Similarly, in Canada, 53% say they would need to use credit or borrow from friends and family if faced with an unexpected $1,000 expense.[4] This is very risky, especially since many don't have adequate health insurance and are an injury away from bankruptcy. All it would take is one major unexpected repair bill for their car and they can find themselves with no means of driving themselves to work, potentially losing their jobs and spiraling into debt. One of the best ways to build financial resilience is having a healthy emergency fund. There are different guidelines for how much money you should have in your emergency fund, generally anywhere from three to six months' living expenses.

If you are a low-income earner and are struggling to make ends meet, it might be difficult to save 3 to 6 months' living expenses. Based on a 2019 study, a realistic amount of savings for low-income households is to have $2,500 in savings to help stay out of financial hardship.[5] However, if your expenses are higher and you rely exclusively on a single income source that could easily be lost during a recession, you should aim for higher savings. If you do lose your job, how long do you think it will take to find another one?

---

[3]McGrath, M., 63% Of Americans Don't Have Enough Savings To Cover A $500 Emergency, *Forbes*, Jan. 6, 2016, Available at: https://www.forbes.com/sites/maggiemcgrath/2016/01/06/63-of-americans-dont-have-enough-savings-to-cover-a-500-emergency/?sh=127333c64e0d

[4]CIBC, Why is it so hard to save? Most Canadians say they 'need to save more,' but aren't making it a priority: CIBC Poll. *Newswire.ca*, Nov. 21, 2017.

[5]Elkins, K., Economists say this is the minimum amount of money you need in an emergency fund, *CNBC make it*, Oct. 18, 2019, Available at: www.cnbc.com/2019/10/18/minimum-amount-of-money-you-need-in-an-emergency-fund.html

You should keep most of this money in high-yield savings accounts that pay higher interest rates, provided that you can have quick access to it. It could also be a good idea to have a minimum amount of cash-on-hand if there is an emergency/disaster that prevents you from accessing your bank. Some people go so far in making sure they have some gold at home if a societal collapse leads to the devaluation of paper money. The war in Ukraine in 2022 and the devaluation of the Russian currency saw people turning to crypto-currency as a backup.

Even though getting into debt can hurt your resilience, having access to credit during an emergency could be beneficial. In case of an emergency, a line of credit can be accessed on short notice. However, it is best to apply for a line of credit when you have a job and your finances are in order. Alternatively, you can apply for a low-interest credit card. These should be used only during emergencies and the balance should be paid off as quickly as possible.

Savings for Future Expenses: An emergency fund can protect you against unexpected expenses and loss of income. You should not use the emergency funds for planned expenses such as vacations and renovations. If you know that you have larger expenses that will be happening in the coming years (buying a new car, kids' education, renovations, new house, etc.) you can begin saving for those expenses now. That way, you don't have to go into debt, or you can at least reduce your debt load when the time comes. If you have these savings, they could also be used as a backup if your emergency fund is exhausted.

It might be useful to invest these savings where you will earn higher returns and where you might not have immediate access to your money, which can help stop you from using it for other short-term needs or wants.

Retirement Savings: In addition to having savings for an emergency fund and future expenses, you need to save for retirement. Looking at statistics, around two-thirds of North American's are worried that they are not saving enough for retirement. This leads many people to push their retirement age so that they can have more time to save and reduce the length of retirement they need to save for. The earlier you can start investing for retirement, the lower the percentage of your income that needs to be devoted to this. However, it can be hard to start saving early in life since you often have student debts, you are earning less than what you will in the future, yet your expenses can be high if you are hoping to buy a house and start a family. If you learn to live frugally with relatively low annual expenses and that you can carry on living that way in retirement, you will need to save less.

Deciding how much money to devote to other savings and retirement while considering your debt level and other financial obligations can be challenging. It is a good idea to invest some money and hire a financial planner who can evaluate your whole financial situation, not just someone that is in the business of selling mutual funds for retirement savings.

Once you think of action items and goals that you would be interested in doing, you can include them in your Personal Resilience Worksheet, which you can either download electronically at www.remicharron.ca/book or use the sheets provided in the Appendix. The following is an example of a plan to build your savings.

Savings plan	
Start this year	Review financial situation with our financial adviser once per year Grow our emergency fund by $2000 and keep $1000 in cash in a safe.
Start next year	Invest my raise into boosting my retirement savings Grow our emergency fund by another $2000 Save for a big canoe camping trip with the family
Medium term, years 3–10	Keep meeting our financial adviser once per year and making recommended adjustments. Look at increasing percent savings needed for retirement. Start saving for trading in our Toyota Echo for an electric car Save for an annual vacation
Long-term, over 10 years	Same as above. Think about saving money to help kids pay for a down-payment for a house.

10.2 Debt Repayment

> *"Annual income, 20 pounds; annual expenditure, 19 pounds; result, happiness. Annual income, 20 pounds; annual expenditure, 21 pounds; result, misery."*
>
> – Charles Dickens

With so many people living paycheck to paycheck, it is no wonder that consumer debt keeps increasing. All it takes is one unexpected expense that you can't pay for and it goes on the credit card. It would be best if you tried to pay it off little by little with whatever spare money you do have. With an average annual interest rate of 18%, credit card debt is expensive to carry, so any extra money you had before is now reduced since you now need to cover minimum credit card payments. When the next emergency happens, you have even less financial capacity to handle it.

Even with people that can afford their debt, simply carrying debt can have an emotional and psychological toll, causing depression and anxiety, resentment, denial, stress, anger, fear, regret, shame and embarrassment.[6] People struggling to pay off their debts are more than twice as likely to suffer from mental health issues. The stress and anxiety about debt will often start to impact other areas of your life. Sonya Britt, assistant professor of family studies at Kansas State University, found that having arguments about money is the top predictor of divorce.

I would argue that paying off high-interest non-mortgage debt is more important than having an emergency fund. There is no use having money in the bank earning close to no interest when you are carrying high-interest debt. If you put your $2,000 emergency fund towards your credit card bill, this will give you an extra $2,000 of room on your credit card in case of emergencies. However, it is crucial to stress that this is only if you are in a situation where you are carrying excess debt. Don't plan to use debt as your emergency fund since that is not a good position to be in. Your debt levels can quickly spiral out of control.

Pay Off Non-Mortgage Debt: Financial resiliency is increased when people live below their means and spend less than they earn and stay out of debt. However, you may find yourself in a position where you have student loans, car loans, credit card debt, etc. The more debt you have, the less able you are to cope with a loss of income. You also end up spending a lot of your hard-earned income on interest, which does not give you anything. If you have debt, you should try to pay off the highest interest loans as quickly as possible. If you find yourself in a lot of debt and find it hard to make ends meet, it would be a good idea to consult with a credit counselor to develop a plan on how best to get out of debt.

---

[6]Kuchar, K., The Emotional Effects of Debt, *The Simple Dollar*, Oct. 28, 2019, Available at: www.thesimpledollar.com/credit/manage-debt/the-emotional-effects-of-debt/?/186

Mortgage Debt: Mortgage debt is a bit different from consumer debt as the interest rates are usually fixed at a lower rate for a longer period of time. It is an investment that can help reduce your living costs later in life. However, buying more house than you can afford is never a good idea. If a high portion of your income goes to mortgage expenses, property taxes, heating costs and condo fees, you are house poor. Real estate prices in some areas are so high that there is no way that an average family can afford to buy a home without getting into too much debt.

Banks will usually lend more money than is prudent in terms of resilience. Their lending decisions are usually based on debt to income ratios that you should not go above. You should typically stick below these ratios. However, you can determine what you can afford if you have a reasonable budget that includes your other savings goals. If buying is too expensive, renting would be a better option than getting into too much debt, provided that you can find a reasonably priced rental that does not eat up too much of your income. Renting can be more resilient in the event of a downturn as you will be better able to relocate somewhere if needed to find a job.

If you do buy a house, it is a good idea to save up for a reasonably sized down payment. Otherwise, if you end up buying a home with a low downpayment and a recession hits shortly after, you can find yourself with an underwater mortgage, where your mortgage is worth more than your house. If you need to move (e.g. you lose your job, get divorced, etc.), you will owe more than your home is worth and would need to pay the difference before you move. When it is time to renew your mortgage, your bank could increase your interest rate or might ask you to pay the difference between what you owe and the reduced property value.

Once you think of action items and goals that you would be interested in doing, you can include them in your Personal Resilience Worksheet, which you can either download electronically at www.remicharron.ca/book or use the sheets provided in the Appendix. The following is an example of a plan for debt repayment.

Debt repayment	
Start this year	Budget to pay off our line of credit in 12 months.
Start next year	Change mortgage payment from monthly to bi-weekly to pay it off faster.
Medium term, years 3–10	Increase mortgage payments by 5%
Long-term, over 10 years	Put any bonuses towards the mortgage until it is paid off.

10.3 Income Generation

> *"It's not how much money you make, but how much money you keep, how hard it works for you and how many generations you keep it for."*
> – Robert Kiyosaki

Savings and debt repayment can help you be more resilient in the event that you lose your job or primary source of income. You can also devote some time and energy to make your income generation more resilient so that there is less risk that you will lose your income for an extended period. This section looks at increasing your worth at work, getting a second job or building up a passive-income stream.

Increase Your Worth: As discussed in the Learning and Education chapter, learning new skills can help you keep your job or find a new one more quickly if needed. Learning new skills and getting new certifications can help boost income by getting you raises at your current job or at least making you less expendable. Your income is probably your single most significant asset, so investing in yourself is a good investment to make.

Get a Side Job: Getting a part-time job or a side-gig in the new gig economy can both boost your income and give you some income to fall back on if you lose your primary job. Think about your interests and what you enjoy doing and consider how you could earn money. Although getting a second job can help make your income generation more resilient, you need to consider how the additional time commitment will impact the rest of your life. If it takes away the time you had to do exercise and stay healthy or reduce the time you have to spend with your friends and family, impacting your connection with your kids, you need to carefully weigh the benefits and pitfalls of working longer hours.

Build More Passive Income: Passive income is earnings you make that require little to no effort to earn. If you own rental properties managed by a property manager, you can earn a steady stream of income with little involvement. Investing in high dividend stocks and bonds can be another approach. Remember to consider impact investing where you purposefully avoid stocks that support fossil fuel companies and other unsustainable industries. It could be that you could earn royalties from a patent, song or book you wrote in the past. The great thing is that passive income keeps coming in regardless of whether you keep working or not. One investment goal that some people have is to "get out of the rat-race", which essentially means that you earn enough passive income to cover all of your living expenses, meaning you no longer need to work.

Getting out of the rat-race is a central point made by author Robert Kiyosaki in his best-selling book Rich Dad Poor Dad.[7] He even created a Ratrace board game that helps teach the concept in a fun way. Although over-priced, the game is good at teaching some financial concepts to kids. I found a $5 second-hand set at Value Village and have played it often with my kids. I read the book in my early twenties and it resonated with me. It led me to buy a four-plex when I was doing my Ph.D. in Montreal. It was a good investment, giving me a cheap place to live while paying off a mortgage. However, when I moved to Kingston, Ontario, which was 2 hours away, I decided to sell it. I found the stress of having a large mortgage and dealing with tenants and repairs too much. If I had looked at getting a property management company to look after it, it would likely have been a profitable investment decision. Although I didn't lose money when I sold it, Montreal's property prices have since skyrocketed. Hindsight is always 20/20 when looking at investment opportunities that we do not capitalize on.

Once you think of action items and goals that you would be interested in doing, you can include them in your Personal Resilience Worksheet, which you can either download electronically at www.remicharron.ca/book or use the sheets provided in the Appendix. The following is an example of a plan for income generation.

Income generation	
Start this year	Review with my financial planner the potential of directing part of my retirement savings towards investments that generate a cash flow.
Start next year	Take python programming course that will improve my worth at my current job by helping me do research related to artificial intelligence, which would also increase my opportunities for a potential future job switch.
Medium term, years 3–10	Continue pursuing professional development opportunities at work. Get a job as an adjunct professor and teach one course/year
Long-term, over 10 years	As above and write a book

[7]Kiyosaki, R., Rich Dad Poor Dad: What the Rich Teach Their Kids About Money That the Poor and Middle Class Do Not!, *Plata Publishing, 2nd edition*, 2017.

10.4 Protecting Your Assets

"Whatever excuses you may have for not buying life insurance now will only sound ridiculous to your widow."
— Unknown Author

If you are looking at being more resilient, it might not be too hard to convince you about the value of insurance. Having insurance can be essential in being able to bounce back quickly after a disaster. You should also consider making your family more resilient in the event that you die. This section looks at insurance and wills.

Protect Yourself with Insurance: Getting insurance is all about risk management and if you manage your risks, you are more resilient. Having enough life insurance for you (and your spouse) can ensure that your family will be resilient in the event of an unfortunate death. Having sufficient auto insurance can not only make sure that you can pay for a new car in the event of an accident but that you also have enough liability coverage to pay if you have injured someone.

In addition to life insurance, you should consider disability insurance if you are not currently covered and if you can afford it. Statistics show that over their working lives, one person in three encounters a serious health issue. If this health issue prevents you from working, you can go from the primary income earner to another dependent. Disability insurance can help provide you with some income if you can no longer work. Check with your current job to see if you are covered. If not, find out how much it would cost to purchase disability insurance.

Flood and fire insurance can be critical to help cover your house if disaster strikes. With climate change, one thing we can expect is more floods and fires. If you are in a high-risk area, it can be that your insurance premiums are increasing as a result of the added climate risk or it might be that your insurance simply won't cover the risks. Many insurance companies will not protect against certain types of flooding that are becoming more common. In Canada alone, some 1.7 million homes are at risk of the types of flooding that aren't included in standard home insurance coverage.[8]

---

[8]Coxon, L., Climate change is upending the home insurance industry. And it's going to cost you, *LowestRates.ca*, Apr. 25, 2019. Available at: www.lowestrates.ca/blog/homes/climate-change-home-insurance-going-to-cost-you

Blair Feltmate, head of the Intact Centre on Climate Adaptation, has been warning about uninsurability since at least 2013. He points out that an increasingly uninsurable housing market will inevitably hurt the mortgage market. He points to 25 cities in Canada that have experienced a lot of flooding, where insurance has been pulled back or is, in many cases, nonexistent. Before buying a house, you should study the climate risks in your region and find out if insurance will cover you. Avoid purchasing properties that will become flooded as sea levels rise. If your insurance doesn't cover some risks, you should consider boosting your savings to cover part of this risk yourself.

Wills and Powers of Attorney: Like insurance, wills and powers of attorney can help protect you and your family in the event of an unfortunate death or illness. Although it is not fun to think about, having a proper will can protect your money and possessions and make sure that your children, spouse and even your pets receive the care you want. A Power of Attorney is important in the event that you become incapacitated and are no longer able to take care of your affairs. Assigning someone with power of attorney allows them to act on your behalf to make legal decisions about your property, finances or medical care. Would you rather choose someone or have the state choose someone for you?

Once you think of action items and goals that you would be interested in doing, you can include them in your Personal Resilience Worksheet, which you can either download electronically at www.remicharron.ca/book or use the sheets provided in the Appendix. The following is an example of a plan for protecting your assets.

Protecting your assets	
Start this year	Update will
	Get a quote for disability insurance
Start next year	Review insurance requirements with a financial planner
Medium term, years 3–10	
Long-term, over 10 years	Renew life insurance, evaluate needed amount with kids now being adults

11

Physical Preparation

"By failing to prepare, you are preparing to fail."
Benjamin Franklin

Abstract

The last category of the personal resilience plan that is covered in the book is about Physical Preparation, which presents the types of things you can buy that can help make you more resilient. This is broken down into three categories: Long-term resilience, Emergency preparedness and Security. This chapter includes items that you can do regardless of if you own your home or not. For those that own their homes, the next chapter covers specific retrofits that could help make their homes more resilient.

Key Words: emergency preparedness, go bag, scenario planning, emergency kit, security.

The last section of the personal resilience plan focuses on Physical Preparation, which is about items you can buy to increase your overall resilience. This seems too often resonate more with people than the other more personal growth elements. It seems much easier to go out and buy things, renovate your home, check items off your list and be relieved that you are now prepared. This seems to fit right in with our current consumerist society. However, consumerism is one of the reasons that humanity faces an uncertain future. The idea that increasing consumption of goods and services is always a desirable goal and that a person's wellbeing and happiness depends on amassing material possessions is fundamentally dangerous. On the other hand, this book's objective is to develop your personal resilience plan, not on changing society to make it more sustainable.

Some amount of resilience planning does depend on getting some things that can help you prepare for a future emergency. I have divided these into three different categories. The first focuses on items you may need to improve your resilience over the long term. The second looks more at emergency supplies that you should have to help you survive after a disaster and the last looks more at improving your overall security.

Nowadays, many of us don't own our own homes where we would be free to retrofit them as we please. Some own condominiums, but with these, you are more limited as an owner as to what you can do to retrofit your home for added resilience. Finally, as a renter, you are often even more limited in terms of your options. In this chapter, I have included things that you can do that would be relevant regardless of the type of home that you live in. The next chapter on Low Carbon Resilience is more specific to homeowners and discusses specific retrofits that they can do to help make their home more resilient.

11.1 Long-term Resilience

"We should remember that good fortune often happens when opportunity meets with preparation."
– Thomas A. Edison

We tend to be a very reactive society. When there is a big snowstorm, everyone runs out to buy snow blowers. When there is a prolonged power outage, generators fly off the shelf. An extended heatwave? Good luck finding an air conditioner. A new global pandemic? It seems toilet paper becomes a critical item that people can't get enough of. These are all events that can be planned for. We know that there will be storms and power outages, raging wildfires, floods and high winds, global pandemics and difficult recessions. Climate change will undoubtedly increase the number and severity of these events. Now is the time to plan for them, so you don't need to be fighting with the hordes of panic buyers, if stores are open at all when disaster strikes.

One item that you can buy that could help you in some disasters are different how-to books. This is one of the first things I bought, including books on homesteading, first aid, butchering meat, tool building, etc. My thoughts were that I didn't necessarily have to learn all these different skills, provided I had some resources that I could learn from, I could get by in an emergency. Nowadays, with YouTube and the Internet, you can find how-to videos on just about anything. However, in some scenarios like prolonged power outages, your trusty side-kick Siri and her wealth of internet knowledge would not be available.

One of the local farmers in Pemberton lost his home to a fire. He had a couple of pigs that he needed to get rid of. To help him out, we bought both pigs. We decided to keep one of the pigs to have piglets and to butcher the other one for our freezer. Since this was last minute, we could not rely on our usual butcher to help, so we needed to do it ourselves. I was happy to have my Basic Butchering book to rely on, although I did also make use of YouTube for those sections that were less clear in the book.

There are other consumer goods that you can get that could potentially be useful in a future emergency. For example, having a minimum set of tools that you can use to make repairs around the house is a good idea. Similarly, a sewing kit with a potential sewing machine could come in handy if you cannot afford or cannot find new clothes. Think of supplies that you would want in different scenarios, such as:

1. A prolonged power outage in the winter, or during a heat wave.
2. A global pandemic where you are not allowed to leave the house for an extended period.
3. Weeks with approaching wildfires and atrocious air quality.
4. A global depression where you and many other people have little to no income and few prospects of getting a job.
5. An earthquake that causes extensive damage in your area.
6. Flooding in your neighborhood for weeks at a time.
7. Tornados, hurricanes and other freak storms cause widespread regional destruction.

These are not pleasant scenarios to think about, but their likelihood will only increase over time. There are some things you could do today to help your future self-cope with disaster.

Earthquakes: Even if you can't fix the structure of your home to make it able to better withstand an earthquake, there are some things you can do to keep yourself safer. Secure suspended ceilings by attaching them to the structure of the building every few feet. Brace water heaters and major appliances with flexible metal armored connectors, especially gas-burning appliances. Do not hang picture frames, mirrors or anything heavy over beds. Avoid hanging light fixtures. Install a special window film on your windows that can protect the inside of your home from shattering glass during an earthquake.

Forest Fire Smoke: Climate change is increasing the frequency and severity of forest fires. It is not uncommon to have thick blankets of smoke covering large areas. If you have poor air quality in your home during these events, you can buy a portable air filter provided that it is rated to filter out smoke (e.g. HEPA). During smoke events, keep your windows closed and the air filter on. Forest fires often happen at the same time as heat waves. If you don't have a cooling system and rely on your windows to keep you cool, you often have a choice to either open your window for passive cooling and have poor air quality or keep your windows closed with your air filter, but be too hot. Consider getting large fans or a portable air-conditioner that you could use to keep yourself cool while having the windows closed for these situations.

Heat Waves: Heat waves are getting hotter and lasting longer than before. Homes that may have been okay without a cooling system can get unbearably hot during these heat events. At my house, we typically keep windows and blinds closed during the day and open them up at night to cool off the house. This even helped keep our home reasonably comfortable when the heat dome was covering the Pacific Northwest in 2021, which brought peak outdoor temperatures of 48°C in the shade on our deck. However, we have two bedrooms in our attic that get way too hot. One has an east-facing window and the other is west. We have a dark metal roof with not much insulation underneath, which really heats up the ceiling. We have a window air-conditioner to help keep the west-facing room cool for the few weeks in the summer that needs it.

If there is a room that gets too hot in your home, you can potentially get a portable air-conditioner to help keep you cool. In addition or alternatively, you can install a ceiling fan or get a larger portable fan to help keep you cool. These are more energy-efficient and can help prevent you from overheating as temperatures increase.

Food Shortages: It does't take a big event to cause panic buying leading to empty grocery shelves. Many natural disasters can shut down the local transportation system required to keep grocery store shelves stocked. Having a stocked freezer and a pantry full of preserves can help you for a while. Having hunting and fishing supplies can be handy in a prolonged emergency where access to food might be limited. As mentioned previously, knowing some local farmers can likely help as well.

Power Outages: Everyone should have flashlights or other light sources that can be relied on during power outages. If you have medical equipment that relies on electricity to keep you healthy, you should likely consider getting a battery backup system for power outages. You could maybe also get a small solar charging system to help keep your phone and other devices charged. Some can be small enough that you can lay them out on a balcony or, alternatively, you can take them out to an open area to charge before bringing them back home.

Water Shortages: Extended droughts might limit the amount of water you have available or longer-lasting power outages might affect the local water distribution system such that you have no water coming from your municipality. You might want to have a certain amount of water that you can use in these situations. At the very least, you might want to consider having some containers that could store water if you needed it. I have a hand pump water filter that I use when I go camping. In a pinch, I could use that to help filter water from a nearby stream.

Once you think of action items and goals to pursue, you can include them in your Personal Resilience Worksheet, which you can either download electronically at www.remicharron.ca/book or use the sheets provided in the Appendix. The following is an example of a physical preparation plan for improving your long-term resilience.

Physical preparation – long term resilience	
Start this year	Purchase a few resilience-related how-to books for my book collection. Replenish my stock of candles and matches that we use for lighting during power outages.
Start next year	Get a couple of portable air purifiers with a stock of extra filters to help improve the air quality when there is smoke. Secure appliances to prepare for a potential earthquake.
Medium term, years 3–10	Get a battery backup with a small solar charger to help during power outages. Install a larger pantry in the kitchen area where I could store more preserves and longer-lasting food.
Long-term, over 10 years	Periodically consider the various scenarios and replenish items that could help make me more resilient.

11.2 Emergency Preparedness

> *"By the time you need an emergency kit, it is too late to buy one."*
> – Remi Charron

A disaster can happen at any time. Families should have an emergency preparedness plan and at least 72 hours of food, water, and medication in an emergency kit. There are several companies that sell pre-assembled kits. We bought a pre-assembled emergency kit with sufficient food, water and other supplies for 72 hours for all the members of our family. We can use this at home if needed or take it with us during an emergency evacuation. I also have a kit for myself in the trunk of my car. I stay in an apartment in Vancouver during the week. Vancouver is an area that is at a higher risk of an earthquake. My apartment is old and would likely be uninhabitable after an earthquake. There is a mountain highway with a number of bridges to get back to my home in Pemberton that may not be passable after an earthquake. The emergency kit in the trunk of my car is in a backpack with some supplies that could help me make the 200 km (125 miles) trek on foot from Vancouver to Pemberton if needed. Although I may never need it, it does give me an element of comfort knowing that it is there.

The www.Ready.gov website has a lot of useful information about planning, including a list of supplies to include in your emergency kit. They recommend that you have your kit in a designated place and have it ready to go in case you have to leave your home quickly. They also recommend a kit at work if you need to shelter at work for at least 24 hours, including food, water and other necessities. A kit is also recommended for your car in case you ever get stranded. The kit I have in my car for a potential earthquake has supplies that would also be useful if I ever get stranded on the road, including a first aid kit that I could use if I am the first on the scene of an accident. If you live in a location with cold winters, this kit should include warm clothing. The FEMA checklist consists of the following items:[1]

1. Water, one gallon (3.8 L) of water per person per day for at least three days, for drinking and sanitation.
2. Food, at least a 3-day supply of non-perishable food.
3. Battery-powered or hand-crank radio and an NOAA Weather Radio with tone alert and extra batteries for both.
4. Flashlight and extra batteries.
5. First aid kit.
6. Whistle to signal for help.
7. Dust mask, to help filter contaminated air and plastic sheeting and duct tape to shelter-in-place.
8. Moist towelettes, garbage bags and plastic ties for personal sanitation.
9. Wrench or pliers to turn off utilities.
10. Can opener for food (if kit contains canned food).
11. Local maps.
12. Prescription medications and glasses.
13. Infant formula and diapers.
14. Pet food and extra water for your pet.
15. Important family documents such as copies of insurance policies, identification and bank account records in a waterproof, portable container.
16. Cash or traveler's checks and change.
17. Emergency reference material such as a first aid book or information.
18. Sleeping bag or warm blanket for each person. Consider additional bedding if you live in a cold-weather climate.
19. Complete change of clothing, including a long-sleeved shirt, long pants and sturdy shoes. Consider additional clothing if you live in a cold-weather climate.

---

[1]FEMA, Emergency Supply List, *ready.gov*, Feb. 2021, Available at: www.ready.gov/sites/default/files/2021-02/ready_checklist.pdf

20. Household chlorine bleach and medicine dropper – When diluted nine parts water to one part bleach, bleach can be used as a disinfectant. Or, in an emergency, you can use it to treat water by using 16 drops of regular household liquid bleach per gallon (3.8 L) of water. Do not use scented, color safe or bleaches with added cleaners.
21. Fire extinguisher.
22. Matches in a waterproof container.
23. Feminine supplies and personal hygiene items.
24. Mess kits, paper cups, plates and plastic utensils, paper towels.
25. Paper and pencil.
26. Books, games, puzzles or other activities for children.

You can think about other potential items that you would want in case of an emergency. For example, you might want a tent if you are displaced from your house but do not want to go to a shelter. Another idea would be to have an extra propane tank if you need to use your barbeque for cooking over an extended period. You could have a certain amount of fuel at home that you could use to get to a destination or at least have a jerry can that you can use to store extra fuel if needed. If an earthquake, tornado or hurricane has damaged the local streets, you might want to make sure you have a bike that you could use to get around. Similarly, if you live in an area prone to flooding, you might want to have a canoe or boat that you could get around with. Think of what else you may need in terms of food, water, health, shelter, transportation and energy.

Once you think of action items and goals that you would be interested in doing, you can include them in your Personal Resilience Worksheet, which you can either download electronically at www.remicharron.ca/book or use the sheets provided in the Appendix. The following is an example of a physical preparation plan for emergency preparedness.

Physical preparation – emergency preparedness	
Start this year	Buy an emergency kit for the car and the office
	Update the emergency kit at home with the FEMA checklist
Start next year	Develop an emergency plan with the family
	Keep the pantry stocked with food.
Medium term, years 3–10	Check and replace batteries in emergency kits
Long-term, over 10 years	

11.3 Security

"If you want total security, go to prison. There you're fed, clothed, given medical care and so on. The only thing lacking...is freedom."
— Dwight D. Eisenhower

For some people, the top of their emergency preparedness equipment includes an arsenal of weapons to protect themselves from the invading hordes. They fear that the unprepared will come to take all of their emergency supplies. It would be a good idea to not flaunt your supplies when others are lacking. However, in a major crisis with severe shortages, if you live in a fortified bunker with lots of supplies and are surrounded by thousands of people that don't, weapons would likely not be enough to protect you. Although movies and books certainly dream up scenarios where having firearms for protection might be needed, in most instances, weapons can be a liability. They can lead to accidental injuries or be used purposefully against you.

In the US, more than 75% of first and second graders know where their parents keep their firearms, and 36% admitted handling the weapons, contradicting their parents' reports.[2] Even in Canada, when I was no older than 4 or 5, I remember finding my dad's hunting bullets (22 calibers), and I was so curious as to what was inside that I tried many things to open them. This included biting down on them and trying to squish them with my dresser. I still cringe at the thought of how stupid that was. Another risk with having a gun in the house is that it increases the likelihood that someone will use it to commit suicide. Over 80% of child firearm suicides involved a firearm belonging to a family member.[3]

For these and other reasons, I would not recommend getting firearms for protection. However, to increase my resilience, I took a hunting course that included getting both hunting and a firearms license. I got a rifle for hunting partridge around the house, which I store under lock and key. There are other things that you could do to increase your security, including:

---

[2] Firearms – Injury Statistics and Incidence Rates, *Stanford Children's Health*, Available at: www.stanfordchildrens.org/en/topic/default?id=firearms--injury-statistics-and-incidence-rat es-90-P02982

[3] Firearm Suicide in the United States, *Every Town Research*, Aug. 30, 2019, Available at: https://everytownresearch.org/report/firearm-suicide-in-the-united-states/

1. You can install a security system in your house.
2. Install Wi-Fi-enabled cameras that allow you to look at different parts of your house on the internet.
3. Get a safe or fireproof box where you keep your valuables and important papers.
4. Install good quality locks on all windows and doors.
5. Install flood lights that turn on automatically with movement.
6. Make sure that your walkways are sufficiently lit-up so that no one can be hiding there when it's dark.
7. Plant thorny shrubs or roses around your property to discourage people from sneaking through.

Nowadays, with our ever-growing online presence, if there is one area where most of us can increase our security, it is online. You shouldn't take shortcuts when it comes to passwords, such as using easy passwords or writing passwords down in a file on your computer. If it is getting too hard for you to remember your passwords, you could get a Password Manager that could help update and manage all your passwords. All you need is to remember the one password for the Password Manager. Some Password Managers can also work with biometrics so that all you need is your fingerprint to log in to anything. We also need a dependable antivirus system with strong firewalls to help protect our computers from hackers.

Once you think of action items and goals that you would be interested in doing, you can include them in your Personal Resilience Worksheet, which you can either download electronically at www.remicharron.ca/book or use the sheets provided in the Appendix. The following is an example of a physical preparation plan to increase your security.

Physical preparation – security	
Start this year	Update firewall on computers. Install a couple of cameras around the farm.
Start next year	Re-visit and change all of my passwords. Get a copy of my credit report to check for anomalies.
Medium term, years 3–10	Get a Password Manager with a biometric scanner to help keep all my passwords secure.
Long-term, over 10 years	Keep abreast of emerging technologies and threats, and follow recommended strategies to keep my online information secure.

12

Low Carbon Resilient Home Retrofits

*"The home to everyone is to him his castle and fortress, as well for his
defence against injury and violence, as for his repose."*
– Edward Coke

Abstract

The last chapter discussing specifics of the personal resilience plan provides
guidance to homeowners on how they can make their homes more resilient.
Given the likelihood that homeowners will also want to reduce the
greenhouse gas emissions from their homes, it is recommended that
resilience and carbon reduction retrofits be considered at the same time.
Starting with an energy and resiliency audit of your home can help you map
out a series of retrofits that will transform your home for the better.

Key Words: low-carbon, net-zero carbon, energy efficiency, retrofits, energy
audit, resiliency audit.

To achieve their greenhouse gas reduction targets, governments around the
world need to implement aggressive retrofit policies to encourage or likely
force people to retrofit their homes. In British Columbia (BC), the
government is aiming for a roughly 60% reduction in emissions from
buildings before 2030. Yet every year, there are roughly 40,000 new homes
and other buildings built in the province, most of which are still relying on
natural gas heating and hot water systems.

The main strategy to decarbonize our buildings involves coupling energy efficiency with fuel-switching natural gas systems to high-efficiency electric heat pumps and induction stoves. Although the solution is relatively easy, there have been many delays in implementing policies to encourage this to happen. There is some aggressive lobbying from natural gas companies that are fighting to keep their businesses profitable. In 2021, news came out that natural gas utilities had actually formed a consortium in an effort to block electrification[1].

These efforts have only been delaying the policies. At some point, governments will have no choice but to mandate aggressive retrofit policies to cut greenhouse gas emissions. The longer they delay, the more aggressive those cuts will need to be. All this to say is that if you own your home, whether you want to or not, in the not too distant future you will likely be required or strongly encouraged, to retrofit your home to reduce its greenhouse gas emissions.

This book is about becoming more resilient, so why am I talking about GHG reductions? There are various retrofits that you can do that will make your home more resilient. There are even some retrofits you can do that will both make your home more resilient and reduce its greenhouse gas emissions. However, the most cost-effective way to take your home from its existing state to a low-carbon resilient home is if you plan on implementing both resilience and carbon reduction measures in a coordinated approach. This chapter will discuss how you can retrofit your home to make it a resilient, more comfortable and healthier home to live in, all while reducing your impact on the climate.

12.1 Overall Retrofit Plan

When it comes to retrofitting your home to make it low to net-zero carbon, it can be relatively easy, depending on a number of factors. It won't necessarily be cheap, but once you're done, you can end up with a more

---

[1]Klein, S., Time to stop playing nice with fossil fuel companies, Canada's National Observer, May 17, 2021. Available at: www.nationalobserver.com/2021/05/17/opinion/ti me-stop-playing-nice-fossil-fuel-companies-blocking-climate-action (Accessed on: Dec. 30, 21).

comfortable, healthier home to live in. However, if the work is done by unqualified professionals, in an uncoordinated way, you can end up with an uncomfortably hot house that has durability issues and costs more to operate than what you started with. Make sure to take the time to check on the experience of the people you hire, ask to speak to a previous clients to make sure they did good quality work. You likely take the time to read reviews on Amazon before buying something small, why wouldn't you take the extra effort to do the same when you hire someone for something much more important?

It is important to get a preliminary audit done on your house by professionals that can give you advice that would be specific to your home. The odds are that it will be easier to find someone that has expertise on the energy efficiency side, but they may not be able to provide too much advice on improving your resilience. That is because energy efficiency programs have been encouraging energy audits and retrofits for decades, whereas resilience is still fairly new. Many governments offer grants and incentives for both energy audits and actual retrofits. Following a bad year in 2021 with many forest fires, intense heatwaves and floods, the Canadian government has indicated that they would integrate resiliency within their energy audit process as well as started providing incentives for certain upgrades to boost resilience. Check with your local governments if any such incentives exist where you live.

The goal of the energy and resiliency audit of your home would be to map out a plan to take your home from its current state to a low-carbon resilient home. You can do it all in one big retrofit if you have the time and money or you can plan to implement work over a number of years. If you start with the end goal in mind, you can better time which elements of the home should be retrofitted first. For example, if you are planning on redoing the siding on your home in the next few years, but were planning to upgrade your heating system first, it may be a good idea to switch the order and add insulation to your home at the same time as you redo the siding. After you add insulation, you can get a smaller heating system that will save you money and will work better than if you had sized your heating system when you had less insulation.

By planning both resilience and GHG reduction retrofits together, you can make better-informed choices that will prevent you from having to repeat work. If we take the siding example again, by factoring in resilience, you could choose a siding material that would be non-combustible to protect against forest fires and harder so that it can better withstand hail. While the siding is off you can do some air sealing on your home which can reduce energy consumption and help keep wildfire smoke outside. If you had just changed your siding with the cheapest option, you wouldn't have the added advantage of energy savings and increased resilience. The cheaper option would likely have ended up costing you more in the long run.

As mentioned, there are fewer professionals that are aware of all the retrofit options that you can do to improve the resilience of your home. Luckily there is a steady stream of information that is being developed to educate homeowners and building professionals on how they can improve resilience. In many cases, the insurance industry is helping fund some of these initiatives as they are trying to keep their losses from becoming unmanageable as climate change gets worse. For example, the Institute of Catastrophic Loss Reduction (ICLR) in Canada has produced a series of homeowner guides on things that can be done to improve your home's resilience. Including how to better withstand severe wind, flooding, fires, snow, hail, earthquakes and extreme heat. I recommend you take the time to review these different guides.[2] In the next sections, I will outline some retrofit considerations for different house systems that can help both increase your resilience and lower your overall carbon footprint.

12.2 Exterior Walls and Windows

As described in the example, if you are redoing the exterior siding or cladding of your home, it would be a good time to add exterior insulation to improve the energy efficiency of your home. For a retrofit in cold climates, it might be a good idea to consider using Rockwool insulation as it can breathe. If your home had an interior vapor barrier to protect it from moisture, installing foam-based insulation products like Styrofoam can cause some moisture problems if water somehow finds its way into your wall. Another advantage of Rockwool insulation is that it is non-combustible and when combined with a non-combustible cladding such as stucco, metal, brick, concrete or fiber cement siding, it can better protect your home from

---

[2]ICLR, "Protect your home from" Homeowner Booklets, Institute of Catastrophic Loss Reduction, Available at: www.iclr.org/homeowner/

forest fires. These harder cladding materials can also help minimize major siding damage in high-risk hail zones compared to vinyl or aluminum siding.

As mentioned, when your siding or cladding is being replaced, it is a good time to do some air-sealing so that your home becomes more airtight. However, if your home does not have a ventilation system installed that brings fresh air into your home, you can end up with some indoor air quality and moisture problems. All those cracks were leaking energy, but they were also bringing in fresh air into your home. The solution is to add a ventilation system that will bring in the fresh air in a controlled approach. This is discussed in the upcoming section on HVAC upgrades.

When redoing your exterior siding, it would also be a good time to consider adding exterior shading devices to your windows. Exterior shading devices are much more effective than internal shades because they reflect solar energy away from the home before it goes through your windows. They can reduce the risk of overheating by more than 90% compared to a reduction of only 11 to 30% for internal reflective blinds.[3] Exterior shading can help reduce cooling energy consumption and can help manage overheating in the event that you don't have a cooling system or that it is not functioning because of a power outage. Storm shutters could be used as exterior shading, and also help protect windows from airborne debris during intense windstorms.

Managing the solar heat gain that goes through your windows is one of the best ways to prevent overheating. This can be done by shading, by adding certain window films to your existing windows or by replacing your windows with ones that allow less solar radiation through. If you are replacing your windows, triple-glazed windows are a good choice in terms of energy efficiency in colder climates. They are also generally quieter than your standard double-glazed windows. The solar heat gain coefficient is a rating that provides the percentage of solar energy that will go through the window opening. A higher value will reduce your heating energy use in winter, but increase your cooling energy use and your risk of overheating in the summer. I would opt for a lower value (<0.3) to help reduce the risk of

---

[3] A. Laouadi, M. Bartko, A. Gaur and M. Lacasse, "Climate resilience buildings: guideline for management of overheating risk in residential buildings," National Research Council of Canada. Construction, Ottawa, 2021.

overheating, especially if the window doesn't have effective exterior shading. However, if the value is too low, it can make your home feel darker.

Other attributes to consider when buying windows would be to consider windows with tempered glass for protection during earthquakes, hailstorms, and windstorms. Tempered glass is also more heat resistant in the event of a fire. If you are not replacing windows, there are certain window films that can be added on to help improve their performance. Some films can help reduce solar gains, whereas others are designed to make them stronger against hail or windblown debris. Some films can also help prevent your window from shattering in the event of an earthquake. Many of these window films can also filter out UV light to protect the interior finishing's in your home.

If you are working on the exterior wall from the inside of the house, it is also a good time to do some air sealing and fix any insulation that may have degraded. It can also be a good time to help improve the ability of your home to withstand an earthquake. As explained in the ICLR Protect Your Home from Earthquakes homeowner guide[4], there are some relatively simple strategies that can be used to reinforce your home. You can hire a professional to evaluate and strengthen the connection between the foundation and walls, between walls and floors and between the walls and roof (see Figure 12.1), as well as reinforce and brace masonry chimneys.

12.3 Roof and Attic

Adding attic insulation is typically a low-cost measure to save energy. Before adding insulation, it would be a good time to make sure that there is good air sealing between your ceiling and your attic to prevent moisture-rich air from your home to make its way to your attic. If you have any exposed pipes in your attic, it would be good to insulate these to protect against freezing. You can also apply heat tape or cables to exposed pipe sections.

To protect against high winds, you can strengthen the fastening of the roof sheathing to the roof structure by increasing the length and quantity of nails/screws, as recommended by guidelines. You can also add hurricane ties

---

[4]Protect your home from Earthquakes, Institute of Catastrophic Loss Reduction, 2016.

to better connect your roof trusses to the house. To protect against forest fires, the roofing should have a Class A non combustibility rating for the best

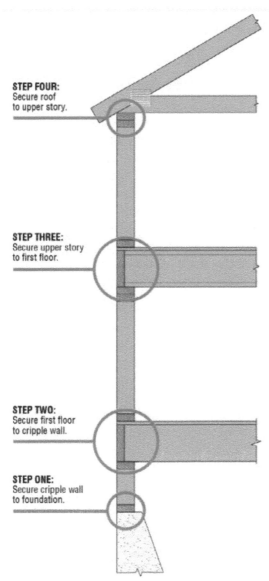

STEP FOUR:
Secure roof
to upper story.

STEP THREE:
Secure upper story
to first floor.

STEP TWO:
Secure first floor
to cripple wall.

STEP ONE:
Secure cripple wall
to foundation.

Figure 12.1 Reinforcing the connections in your home can help reduce the damage your home will experience in an earthquake (Source: ICLR)

protection. All vents should be fire-rated or screened with three mil metal mesh. Soffits and fascia should be properly fitted to protect rafters from embers. Class 4 rated roofing can protect against hail. Before adding your roof cover, you can add a waterproof underlayment as a second layer of protection in the event that your roof cover is damaged.

12.4 HVAC and Plumbing Systems

As discussed, one of the best ways to reduce your greenhouse gas emissions is to fuel switch from natural gas burning appliances to high-efficiency heat pump systems. In areas where the electricity is produced by coal or other fossil fuel sources, the immediate GHG benefits might be less. However, most utilities are also transitioning from fossil fuel to renewable energy reducing the GHG emissions associated with electricity in the process. You can also help offset your electrical emissions by installing a solar photovoltaic (PV) system in your house. If your house has a basement that is liable to be flooded in the future, it would be a good idea to install your HVAC systems on the main or upper level, if possible.

One of the cheapest ways to save energy is by air sealing your home. However, if you do this, you need to make sure that you have an adequate ventilation system to bring in the fresh air. A heat-recovery ventilator (HRV) is a ventilation system that will not only bring in the fresh air but will also capture some of the heat from the air that you are exhausting outside as shown in Figure 12.2. Some HRVs also come with the ability to add air filtration, which can help filter out wildfire smoke. The filter should be rated to be able to remove smoke (e.g. minimum MERV 13 in North America or F7 in Europe).

To protect your systems from earthquakes, you can replace all rigid water and/ or natural gas appliance connections with flexible connections, flexible armored connectors and shut-off valves. To protect against flooding, get a plumber to determine whether a sewer backwater valve should be installed (in accordance with municipal guidelines and possible subsidy programs). If your basement has a sumppump, installing a second backup sumppump, both with backup power, can improve the resilience of your system in the event of a flood with a power outage.

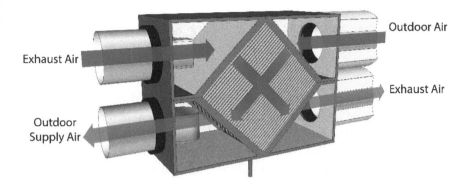

Figure 12.2   Heat-recovery ventilator captures heat from exhaust air to preheat your incoming outdoor air. [5] Copyright BC Housing. Reprinted with permission.

12.5 Solar and Backup Power

Having backup power can improve your resilience to power outages. Identify the appropriate size, form and location of backup power for your needs. Solar photovoltaic systems are becoming cheaper every year and can provide some backup electricity during power outages if designed with a backup battery system. If you install a solar system with no battery backup, you will not be able to access the solar-generated electricity during a power outage. Backup systems are generally not designed to be able to replace all of your electrical needs. You would need to think about what electrical circuits you still want to be powered after a power outage. Having your fridge and freezer supplied by backup power would be a good idea to prevent food spoilage during a more prolonged power outage.

If you have a large solar system, there are times of the year when you would likely have excess power, which you could use to charge an electric car or other loads. There are some electric vehicles that can also be used as a backup battery in the event of a power outage such as the Ford F-150 Lightning. In the winter, your energy demands can be higher, while solar

---

[5]BC Housing, Heat Recovery Ventilation Important Considerations for Builders and Designers, *Builder Insight Technical Bulletin No 14*, Mar. 2016, Available at: www.bcho using.org/publications/Builder-Insight-14-Heat-Recovery-Ventilation.pdf

generation is lower. You can install a nonelectric standby stove or heater for these instances.

12.6 Landscaping

Landscaping can play an important part in reducing the risk that wildfires pose to your home. Read the Fire Smart Homeowner Manual and follow the guidelines.[6] As shown in Figure 12.3, a fire-resistant zone should be located from 5 to 30 feet (1.5 m to 10 m) of the house, where it is free of all materials that could easily ignite from a wildfire. Maintain a 1.5-meter non-combustible surface around your entire home and any attachments, such as decks. Space trees at least 10 feet (3 m) apart and prune all tree branches within 7 feet (2 m) of the ground. Enclose decks and balconies or sheath in the base with fire-resistant material to reduce the risk of sparks and embers igniting your home.

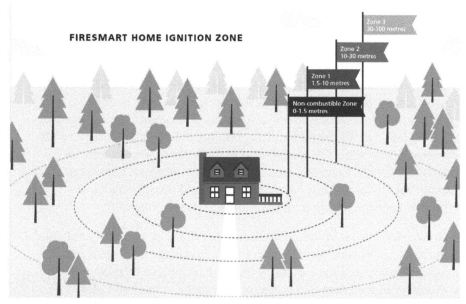

Figure 12.3   Different zones of protection around your home to protect against wildfires [7] (Source: ICLR)

---

[6]FireSmart Begins at Home Manual, *FireSmart Canada*, Nov. 2018.
[7]Protect your home from Wildfire, Institute of Catastrophic Loss Reduction, 2019.

Landscaping can also be used to reduce your risk of overheating. Use shading trees and other plants near south, east and west-facing exposures to minimize solar heat gains through windows while keeping in mind the Fire Smart guidelines.

Extended droughts might limit the amount of water you have available or longer-lasting power outages might affect the local water distribution system such that you have no water coming from your municipality. Building a rainwater capturing system could help provide you with a source of water. This could at least be used for your gardening needs and could also be used as a source of drinking water if you boiled it or had a water purification system.

Landscaping is also an integral part of protecting your home from flooding. Sea levels are expected to rise substantially in the coming decades. The severity of rain events has also been increasing to the point where they can exceed the local storm systems' capacity. Before buying a house, make sure that it sits above the revised future flooding predictions. If you already own your home and it's going to likely be impacted by sea-level rise, you should consider selling before others have the same idea.

You can reduce your vulnerability to flooding in your existing home. Seal cracks in foundation walls and basement floors. Downspouts should flow through extensions directed at least 1.8 meters (6 ft.) away from the home. The flows should be directed over permeable surfaces, such as lawns or gardens. Ensure that your yard slopes away from the house. The ground beside your foundation should be approximately 4 to 6 inches (10 cm to 15 cm) higher than the soil 5 feet (1.5 meters) away as shown in Figure 12.4.

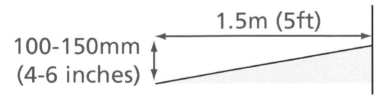

Figure 12.4 Proper yard grading showing a slope away from your house of at least 4 inches (10 cm) of vertical drop for every 5 ft. (1.5 m) of horizontal distance[8] (Source: ICLR).

---

[8]Protect your home from Basement flooding, Institute of Catastrophic Loss Reduction, 2019.

With the information that you learned in this chapter, you can update the Long Term Resilience section of your Personal Resilience Worksheet presented in Chapter 11. I have now covered all the material. The next step is putting the plan into action.

13

Putting Plan to Action

"A good plan today is better than a perfect plan tomorrow."
– George S. Patton

Abstract

The last chapter encourages the reader to start their plan based on what was covered in the book. The book refers to a worksheet that readers can download and fill out as part of their plan. It also has the tables from the workbook at the end of the book for those readers that would rather do it on paper.

Key Words: SMART goals, implementation, personal resilience worksheet, action, roadmap.

Given the current state of the world, there are many reasons why you may want a plan to help you become more resilient. Whether you were caught unprepared for COVID 19 and you want to be ready for the next crisis or are scared of what climate change has in store for us, now is the time to act. However, action without direction won't get you where you want to go. The plan laid out in this book can allow you to take control of your life by helping you draft a roadmap to your more resilient future.

Developing a plan to boost your resilience does not mean you have to drop everything and start preparing for a coming storm. Most of the elements of the plan are things that are already recommended that you do as part of living a happy, fulfilling life; exercise, eat healthily, meditate, keep learning, embrace friends and family, stay out of debt, save for the future and be prepared for emergencies. I took the preparation to the extreme. I quit my job, moved across the country from the city to a farm, from my own house to living with my in-laws, from having a full-time job to being self-employed. It was a very drastic change that turned out to be more difficult than I had envisioned. It will be up to you to decide how much change to commit to.

When planning for future climate change, your specific situation can help inform what you need to prepare for. Are you currently working in an industry that will be impacted by climate change itself (e.g. forestry, agriculture, etc.), or working in an industry that will be impacted by policies implemented to reduce climate change (e.g. oil and gas, automobile industry, etc.)? Do you currently live and potentially own property on the coast where rising sea levels are a serious future threat? Do you live in a house/condo that is already unbearably hot in the summer? Do you have asthma and struggle to breathe with forest fire smoke? Many governments (municipal, provincial/state, national) have released studies that estimate how much their climates will change in the coming decades. Take the time to research reports that would provide predictions relevant to your location. A bit of research and thought given to more specific climate impacts that you will encounter will help make your plan even more relevant.

As you start thinking about activities and filling out the Personal Resilience Worksheet, be realistic in what you include in your plan. Don't overload it with so many things that you give up because you start falling behind on too many items. Also, don't start with too many elements that are outside of your comfort zone. It's okay to push your boundaries by trying something new, but including too many things that are scary to you or just not things that you would normally do, can also lead you to forget about it and retreat to your comfort zone. You want to include S.M.A.R.T. goals in your plans that are Specific, Measurable, Achievable, Relevant and Time-bound:

Specific: Avoid being too general. For example, instead of saying that you want to start exercising, include what type of exercise and how often you want to do it.

Measurable: Write your goal in a way that you can measure success and determine when you have achieved it. This makes a goal more tangible because it provides a way to gauge your progress. For example, you may want to finish a race under a certain time or take a course to achieve a specific designation.

Achievable: Make sure that you set reasonable goals that you will be able to achieve in the timeframe that you have given yourself. Keep in mind that the Resilience Plan presented in this book has a lot of different components. You likely can't achieve everything in the first year. Make sure that you don't include too many things that you won't have time to complete.

Relevant: The goal should have some relevance to making yourself more resilient.

Time-bound: The way the Resilience Plan is structured with items to start this year, next year in the medium term and the long term, which can help you set some time bounds. Having goals with specific deadlines often encourages us to work on them compared to goals that can be done within an open-ended timeframe.

The good news is that your Resilience Plan does not have to be perfect from the start. You can adjust it as needed over time. The bad news is that the plan is not something you develop once and forget about; you should revisit it periodically. The way it is structured is to include items that you will start this year, next year, in the medium term (years 3–10), and long-term (over 10 years). Once your plan is complete, I would suggest doing some tweaks along the way and doing a major review every 2 years. After 2 years, you can essentially redo the plan, focusing on the efforts that you will make in the coming 2 years. Some of the medium-term items can be shifted to activities for next year, and potentially some of the long-term items can move to your medium-term column. This will be a living plan that will evolve as you continue on your journey of becoming more resilient. It should evolve as your life changes with kids, spouse, job, new home, etc.

When looking at specific items to include in your plan, you will find that there are some one-off items and some that are ongoing. For example, completing a specific course or reading a particular book are one-off items; once done, you can check them off your list. Others like becoming a vegetarian or going to the gym three times a week are habits that you want to adopt on an ongoing basis. For ongoing items, include them in each of the time-scales following its first entry. For example, if you begin something next year, include it in the medium and long-term columns. Once you redo your plan after 2 years, items that have truly become habits don't need to go on your plan. On the other hand, if you are still struggling to maintain a habit, I suggest you include it again in your plan to keep the pressure on.

You can do many activities as part of your plan that will count towards multiple categories. Making friends or simply taking the time to deepen your relationship with existing friends will help with community building and be beneficial to your mental health. When you want to learn something new, if you take an in-person class in your local community, you will be learning while growing your community. Similarly, when getting into shape, instead of doing solo activities, being part of a club or team or asking a friend to be your exercise partner will again help you with your community building. By being strategic in developing your plan and including more items that cover

Example activities that increase resilience in more than one category	Fitness and health	Learning and education	Community building	Financial planning	Physical preparation
Join a jogging, exercise or other sports club/team	X		X		
Read an investment book		X		X	
Take a tax preparation course		X	X	X	
Taking any in-person course		X	X		
Buy camping gear and go camping as a family	X	X	X		X
Learn to preserve food (buy equipment & take a course)	X	X	X		X
Learn a new musical instrument (buy and take a course)		X	X		
Buy a house with room for a garden				X	X
Invest in making home more energy efficient				X	X
Invest in a solar photovoltaic system				X	X
Take mediation/yoga class	X	X	X		
Buy and use home exercise equipment	X				X

multiple categories, you can get more accomplished with less effort. The table provides several examples of activities that can fall into multiple categories. When filling out your plan, you can include them in all of the relevant categories in your worksheet.

If you haven't already started, now is the time to fill out your Personal Resilience Worksheet. You can either download an electronic copy at www.remicharron.ca/book or fill out the tables provided in the Appendix. You can tackle it by completing one category at a time over multiple days or you can set aside a day where you brainstorm ideas and fill in the plan. One suggestion that could help you would be to do this with a few friends or family, where you brainstorm together but fill out individual plans. Committing to developing and executing your resilience plans together could be a good bonding experience to strengthen your relationship. It can also help provide you with the needed motivation to put your plan into action. So get to it!

Appendix

Fitness and health	Start this year	Start next year	Medium Term, years 3–10	Long-term, over 10 years
Getting and staying in shape				
Improving your diet				
Health prevention & improvements				
Improving your mood				
Improving your psychological resilience				

Learning and education	Start this year	Start next year	Medium term, years 3–10	Long-term, over 10 years
Children's education				
Career skills				
Skills to increase resilience				
General knowledge building				

Community building	Start this year	Start next year	Medium term, years 3–10	Long-term, over 10 years
Neighbors				
Local community				
Friends and family				

Financial	Start this year	Start next year	Medium term, years 3–10	Long-term, over 10 years
Savings				
Debt Repayment				
Income Generation				
Protecting your assets				

Physical preparation	Start this year	Start next year	Medium term, years 3–10	Long-term, over 10 years
Long-term resilience				
Emergency supplies				
Security				

Index

About the Author

Remi Charron is a professional engineer with a Ph.D. in net-zero energy buildings. He is an Associate Professor and Program Coordinator for a Master's of Energy Management at the New York Institute of Technology's Vancouver campus. He also works as an independent consultant for the provincial (British Columbia) and Canadian governments on projects related to making housing more energy-efficient and resilient. He lives in Pemberton, British Columbia, where he and his wife run a small family farm. The book is premised on a plan he developed to make himself and his family more resilient that started 10 years ago when he quit a government job to move across the country to a family farm.